THE NEW
Art and Science
OF TEACHING

Science

BRETT ERDMANN   STEVEN M. WOOD
TROY GOBBLE   ROBERT J. MARZANO

A joint publication

ASCD   Solution Tree

555 North Morton Street
Bloomington, IN 47404
800.733.6786 (toll free) / 812.336.7700
FAX: 812.336.7790

email: info@SolutionTree.com
SolutionTree.com

Visit **go.SolutionTree.com/instruction** to download the free reproducibles in this book.

Printed in the United States of America

Library of Congress Control Number: 2022026413

Solution Tree
Jeffrey C. Jones, CEO
Edmund M. Ackerman, President

Solution Tree Press
President and Publisher: Douglas M. Rife
Associate Publisher: Sarah Payne-Mills
Managing Production Editor: Kendra Slayton
Editorial Director: Todd Brakke
Art Director: Rian Anderson
Copy Chief: Jessi Finn
Senior Production Editor: Christine Hood
Content Development Specialist: Amy Rubenstein
Proofreader: Elijah Oates
Text Designer: Fabiana Cochran
Cover Designer: Laura Cox
Associate Editor: Sarah Ludwig
Editorial Assistants: Charlotte Jones and Elijah Oates

Acknowledgments

We would like to acknowledge our science teacher colleagues with whom we have worked at Adlai E. Stevenson High School. They embody the spirit of innovation by always seeking new strategies to ensure high levels of science learning for all students. They inspire us every day, and we are grateful to be their partners in this work.

Visit **go.SolutionTree.com/instruction** to download the free reproducibles in this book.

Table of Contents

About the Authors vii

Introduction 1
   The Overall Model 2
   The Need for Subject-Specific Models 6
   About This Book 7

Part I: Feedback 9

1   **Providing and Communicating Clear Learning Goals** 11
   Element 1: Providing Scales and Rubrics 11
   Element 2: Tracking Student Progress 16
   Element 3: Celebrating Success 19
   Summary 22

2   **Using Assessments** 23
   Element 4: Using Informal Assessments of the Whole Class 23
   Element 5: Using Formal Assessments of Individual Students 27
   Summary 30

Part II: Content 31

3   **Conducting Direct Instruction Lessons** 33
   Element 6: Chunking Content 34
   Element 7: Processing Content 38
   Element 8: Recording and Representing Content 44
   Summary 52

4   **Conducting Practicing and Deepening Lessons** 53
   Element 9: Using Structured Practice Sessions 54
   Element 10: Examining Similarities and Differences 59
   Element 11: Examining Errors in Reasoning 62
   Summary 65

5   **Conducting Knowledge Application Lessons** 67
   Element 12: Engaging Students in Cognitively Complex Tasks 67
   Element 13: Providing Resources and Guidance 69
   Element 14: Generating and Defending Claims 72
   Summary 75

6   **Using Strategies That Appear in All Types of Lessons** 77
     Element 15: Previewing Strategies 78
     Element 16: Highlighting Critical Information 80
     Element 17: Reviewing Content 82
     Element 18: Revising Knowledge 85
     Element 19: Reflecting on Learning 88
     Element 20: Assigning Purposeful Homework 90
     Element 21: Elaborating on Information 91
     Element 22: Organizing Students to Interact 93
     Summary 100

Part III: Context 101

7   **Using Engagement Strategies** 103
     Element 23: Noticing and Reacting When Students Are Not Engaged . 104
     Element 24: Increasing Response Rates 105
     Element 25: Using Physical Movement 108
     Element 26: Maintaining a Lively Pace 109
     Element 27: Demonstrating Intensity and Enthusiasm 112
     Element 28: Presenting Unusual Information 114
     Element 29: Using Friendly Controversy 118
     Element 30: Using Academic Games 121
     Element 31: Providing Opportunities for Students to Talk About Themselves. 124
     Element 32: Motivating and Inspiring Students. 127
     Summary 130

8   **Implementing Rules and Procedures and Building Relationships** . . 131
     Element 34: Organizing the Physical Layout of the Classroom . . . 131
     Element 39: Understanding Students' Backgrounds and Interests . . 134
     Summary 139

9   **Developing Expertise** 141
     Step 1: Conduct a Self-Audit 142
     Step 2: Select Goal Elements and Specific Strategies 144
     Step 3: Engage in Deliberate Practice and Track Progress 144
     Step 4: Engage in Continuous Improvement by Planning for Future Growth . 146
     Summary 146

Afterword 147

Appendix A
***The New Art and Science of Teaching* Framework Overview** 149

Appendix B
List of Figures and Tables 163

References and Resources 167

Index 171

About the Authors

Brett Erdmann is assistant director of science at Adlai E. Stevenson High School in Lincolnshire, Illinois. In his role, he continues to teach AP Biology while also supporting teacher development and helping to lead science department initiatives. Brett began his career in medical and laboratory sales but later attended graduate school to allow him to pursue his passion for teaching science. In addition to his assistant director role, Brett has sponsored many school clubs, served as the school's community service coordinator, and developed a summer STEM research program.

Brett earned a bachelor's degree in biology and pre-medicine from Augustana College, a master's degree in secondary education from Roosevelt University, and a master's degree in educational leadership from American College of Education.

Steven M. Wood, PhD, is director of science at Adlai E. Stevenson High School in Lincolnshire, Illinois. In his role, he provides support and leadership for science learning, including the science faculty's professional development. Steve began his career in the healthcare industry but soon realized that his passion was in helping students become better science learners. In addition to his science teaching and leadership roles, Steve has coached many athletic teams and led several international science travel courses. He is a member of many professional organizations.

Steve earned a bachelor's degree in biology education from Taylor University, a master's degree in biology from Northeastern Illinois University, and a doctorate in educational leadership and policy studies from Loyola University Chicago.

Troy Gobble is principal of Adlai E. Stevenson High School in Lincolnshire, Illinois. He previously served as assistant principal for teaching and learning at Stevenson. Troy taught science for eighteen years and served as the science department chair for eight years at Riverside Brookfield High School in Riverside, Illinois.

The United States Department of Education (USDE) describes Stevenson as "the most recognized and celebrated school in America," and Stevenson is one of only three schools to win the USDE National Blue Ribbon Schools Award on four occasions. Stevenson was one of the first comprehensive schools that the USDE designated a New American High School as a model of successful school reform, and it is repeatedly cited as one of America's top high schools and the birthplace of the Professional Learning Communities (PLCs) at Work® process.

Troy holds a master of science in educational administration from Benedictine University, a master of science in natural sciences (physics) from Eastern Illinois University, and a bachelor's degree in secondary science education from the University of Illinois at Urbana-Champaign.

Robert J. Marzano, PhD, is cofounder and chief academic officer of Marzano Resources in Denver, Colorado. During his fifty years in the field of education, he has worked with educators as a speaker and trainer and has authored more than fifty books and two hundred articles on topics such as instruction, assessment, writing and implementing standards, cognition, effective leadership, and school intervention. His books include *The New Art and Science of Teaching*, *Leaders of Learning*, *Making Classroom Assessments Reliable and Valid*, *The Classroom Strategies Series*, *Managing the Inner World of Teaching*, *A Handbook for High Reliability Schools*, *A Handbook for Personalized Competency-Based Education*, and *The Highly Engaged Classroom*. His practical translations of the most current research and theory into classroom strategies are known internationally and are widely practiced by both teachers and administrators.

He received a bachelor's degree from Iona College, a master's degree from Seattle University, and a doctorate from the University of Washington.

To learn more about Dr. Marzano, visit www.marzanoresources.com.

To book Brett Erdmann, Steven M. Wood, Troy Gobble, or Robert J. Marzano for professional development, contact pd@SolutionTree.com.

Introduction

The New Art and Science of Teaching (Marzano, 2017) is a comprehensive model of instruction with a rather long developmental lineage. Specifically, four books spanning two decades precede and inform *The New Art and Science of Teaching* and its use in the field.

1. *Classroom Instruction That Works: Research-Based Strategies for Increasing Student Achievement* (2nd ed.; Dean, Hubbell, Pitler, & Stone, 2012)
2. *Classroom Management That Works: Research-Based Strategies for Every Teacher* (Marzano, Marzano, & Pickering, 2003)
3. *Classroom Assessment and Grading That Work* (Marzano, 2006)
4. *The Art and Science of Teaching: A Comprehensive Framework for Effective Instruction* (Marzano, 2007)

The first three books address specific components of the teaching process, namely instruction, management, and assessment. The final book puts all three components together into a comprehensive model of teaching. It also makes a strong case for the fact that research (in other words, science) must certainly guide good teaching, but teachers must also develop good teaching as art. Even if they use precisely the same instructional strategies, two highly effective teachers will have shaped and adapted those strategies to adhere to their specific personalities, the subject matter they teach, and their students' unique needs. Stated differently, we can never accurately articulate effective teaching as a set of strategies that all science teachers must execute in precisely the same way.

The comprehensive model in the 2017 book *The New Art and Science of Teaching* (Marzano, 2017) reflects a greatly expanded and updated version of *The Art and Science of Teaching* (Marzano, 2007). One of the unique aspects of *The New Art and Science of Teaching* is that it focuses on what happens in the minds of students by taking a student-outcome perspective as the primary influence. Specifically, when science teachers employ instructional strategies, it generates certain mental states and processes in the learner's mind that facilitate student learning. This dynamic represents the major feature of this new model and is depicted in figure I.1.

Source: Marzano, 2017, p. 5.

Figure I.1: The teaching and learning progression.

According to figure I.1 (page 1), the intervening variable between the effective application of an instructional strategy and enhanced student learning is specific mental states and processes in the minds of learners. If teachers do not produce these mental states and processes as a result of employing a given strategy, then that strategy will have little or no effect on students. This implies that teachers should heighten their level of awareness as they use instructional strategies for maximum efficacy in the science classroom.

The Overall Model

The model in *The New Art and Science of Teaching* (Marzano, 2017) is a framework that educators can use to organize the majority (if not all) of the instructional strategies that research and theory identify. It has several parts: three overarching categories, ten design areas, and forty-three specific elements that serve as umbrellas for a host of instructional strategies.

Three Categories

At the highest level of organization, the model offers three overarching categories: feedback, content, and context.

1. *Feedback* refers to the all-important information loop teachers must establish with students so that students know what they should be learning about specific topics and their current level of performance on these topics.
2. *Content* refers to the sequencing and pacing of lessons such that students move smoothly from initial understanding to applying knowledge in new and creative ways.
3. *Context* refers to those strategies that ensure all students meet these psychological needs: engagement, order, a sense of belonging, and high expectations.

Embedded in these three overarching categories are more specific categories of teacher actions (design areas).

Ten Design Areas

In *The New Art and Science of Teaching* model, each of the ten design areas is associated with a specific teacher action, as follows.

1. Providing and communicating clear learning goals
2. Using assessments
3. Conducting direct instruction lessons
4. Conducting practicing and deepening lessons
5. Conducting knowledge application lessons
6. Using strategies that appear in all types of lessons
7. Using engagement strategies
8. Implementing rules and procedures
9. Building relationships
10. Communicating high expectations

Table I.1 shows the ten teacher actions within the three categories and a description of the desirable student mental states and processes for each. For example, when the teacher conducts a direct instruction lesson (the third design area) within the content category, the goal is that students understand which parts of the new core science ideas are important and how the parts all fit together.

Table I.1: Teacher Actions and Student Mental States and Processes

	Teacher Actions	Student Mental States and Processes
Feedback	Providing and Communicating Clear Learning Goals	1. Students understand the progression of knowledge they are expected to master and where they are along that progression.
	Using Assessments	2. Students understand how test scores and grades relate to their status on the progression of knowledge they are expected to master.
Content	Conducting Direct Instruction Lessons	3. When content is new, students understand which parts are important and how the parts fit together.
	Conducting Practicing and Deepening Lessons	4. After teachers present new content, students deepen their understanding and develop fluency in skills and processes.
	Conducting Knowledge Application Lessons	5. After teachers present new content, students generate and defend claims through knowledge application tasks.
	Using Strategies That Appear in All Types of Lessons	6. Students continually integrate new knowledge with old knowledge and revise their understanding accordingly.
Context	Using Engagement Strategies	7. Students are paying attention, energized, intrigued, and inspired.
	Implementing Rules and Procedures	8. Students understand and follow rules and procedures.
	Building Relationships	9. Students feel welcome, accepted, and valued.
	Communicating High Expectations	10. Typically reluctant students feel valued and do not hesitate to interact with the teacher or their peers.

Source: Marzano, 2017, pp. 5–6.

Each of the ten design areas corresponds with a *design question*. These questions help teachers plan units and lessons within those units. Table I.2 shows the design questions that correspond with each design area.

Table I.2: Design Questions

	Design Areas	Design Questions
Feedback	1. Providing and Communicating Clear Learning Goals	How will I communicate clear learning goals that help students understand the progression of knowledge they are expected to master and where they are along that progression?
	2. Using Assessments	How will I design and administer assessments that help students understand how their test scores and grades are related to their status on the progression of knowledge they are expected to master?
Content	3. Conducting Direct Instruction Lessons	When content is new, how will I design and deliver direct instruction lessons that help students understand which parts are important and how the parts fit together?
	4. Conducting Practicing and Deepening Lessons	After presenting content, how will I design and deliver lessons that help students deepen their understanding and develop fluency in skills and processes?
	5. Conducting Knowledge Application Lessons	After presenting content, how will I design and deliver lessons that help students generate and defend claims through knowledge application?

continued →

	Design Areas	Design Questions
Content	6. Using Strategies That Appear in All Types of Lessons	Throughout all types of lessons, what strategies will I use to help students continually integrate new knowledge with old knowledge and revise their understanding accordingly?
Context	7. Using Engagement Strategies	What engagement strategies will I use to help students pay attention, be energized, be intrigued, and be inspired?
	8. Implementing Rules and Procedures	What strategies will I use to help students understand and follow rules and procedures?
	9. Building Relationships	What strategies will I use to help students feel welcome, accepted, and valued?
	10. Communicating High Expectations	What strategies will I use to help typically reluctant students feel valued and comfortable interacting with their peers and me?

Source: Marzano, 2017, pp. 6–7.

Within the ten categories of teaching actions, we have organized sets of strategies in even more fine-grained categories, called *elements*. As teachers think about each design question, they can then consider specific elements with the design area.

Forty-Three Elements

The forty-three elements provide detailed guidance about the nature and purpose of a category of strategies. Table I.3 depicts the full complement of elements that correspond to each design area. For example, we operationally define the category *practicing and deepening lessons* as:

- Using structured practice lessons (element 9)
- Examining similarities and differences (element 10)
- Examining errors in reasoning (element 11)

As a teacher considers how to provide and communicate clear learning goals that help students understand the progression of knowledge he or she expects them to master and where they are along that progression (design question 1), the teacher might think more specifically about providing scales and rubrics, tracking student progress, and celebrating success. These are the elements within the first design area.

Finally, these forty-three elements encompass hundreds of specific instructional strategies. Selected strategies related to science instruction are the focus of this book.

General and Subject-Specific Strategies

At the finest level of detail are more than 330 specific instructional strategies embedded in the forty-three elements. For example, element 24 (increasing response rates) includes the following nine strategies.

1. Random names
2. Hand signals
3. Response cards
4. Response chaining
5. Paired response
6. Choral response
7. Wait time
8. Elaborative interrogation
9. Multiple types of questions

Table I.3: Elements Within the Ten Design Areas

Feedback	Content	Context
Providing and Communicating Clear Learning Goals 1. Providing scales and rubrics 2. Tracking student progress 3. Celebrating success **Using Assessments** 4. Using informal assessments of the whole class 5. Using formal assessments of individual students	**Conducting Direct Instruction Lessons** 6. Chunking content 7. Processing content 8. Recording and representing content **Conducting Practicing and Deepening Lessons** 9. Using structured practice sessions 10. Examining similarities and differences 11. Examining errors in reasoning **Conducting Knowledge Application Lessons** 12. Engaging students in cognitively complex tasks 13. Providing resources and guidance 14. Generating and defending claims **Using Strategies That Appear in All Types of Lessons** 15. Previewing strategies 16. Highlighting critical information 17. Reviewing content 18. Revising knowledge 19. Reflecting on learning 20. Assigning purposeful homework 21. Elaborating on information 22. Organizing students to interact	**Using Engagement Strategies** 23. Noticing and reacting when students are not engaged 24. Increasing response rates 25. Using physical movement 26. Maintaining a lively pace 27. Demonstrating intensity and enthusiasm 28. Presenting unusual information 29. Using friendly controversy 30. Using academic games 31. Providing opportunities for students to talk about themselves 32. Motivating and inspiring students **Implementing Rules and Procedures** 33. Establishing rules and procedures 34. Organizing the physical layout of the classroom 35. Demonstrating withitness 36. Acknowledging adherence to rules and procedures 37. Acknowledging lack of adherence to rules and procedures **Building Relationships** 38. Using verbal and nonverbal behaviors that indicate affection for students 39. Understanding students' backgrounds and interests 40. Displaying objectivity and control **Communicating High Expectations** 41. Demonstrating value and respect for reluctant learners 42. Asking in-depth questions of reluctant learners 43. Probing incorrect answers with reluctant learners

Source: Marzano, 2017, p. 8.

In effect, there are nine distinctive, specific instructional strategies teachers can use to increase students' response rates, supporting the fact that two different teachers could both effectively improve their students' learning by boosting response rates but with very different techniques. Note that throughout the text, we have addressed selected elements—and strategies within elements—that best relate to science instruction.

Therefore, the breadth of this book will not extend to explanations and examples related to science instruction for each of the more than three hundred strategies. For example, this book's exploration of element 24 includes only three of the nine strategies listed here.

Appendix A (page 149) presents an overview of the entire *New Art and Science of Teaching* framework featuring the three overarching categories (feedback, content, context), ten teacher actions, forty-three elements, and accompanying strategies. This can serve as an advance organizer while reading this book.

Some elements draw out the use of the same or similar instructional approaches; for example, the concept of *summary* appears throughout the book, either within a strategy or as a specific name of a strategy. This is because teachers will use summarization strategies differently depending on their particular purposes, as we show in the following examples.

- **Element 7:** In chapter 3 (page 38), "Conducting Direct Instruction Lessons," element 7—processing content—students are asked to assume different viewpoints through perspective analysis and thinking hats (strategies 31 and 32). By comparing and contrasting different viewpoints, students are able to summarize complex arguments.
- **Element 8:** In chapter 3 (page 44), "Conducting Direct Instruction Lessons," element 8—recording and representing content—students are asked to summarize content briefly and quickly to identify critical content and describe how the pieces fit together (strategy 40).

The Need for Subject-Specific Models

General frameworks like *The New Art and Science of Teaching* certainly have their place in a teacher's understanding of effective instruction. However, teachers must adapt those models to specific subject areas to produce the most powerful results. That is what we have attempted to do in this book. Specifically, in the following chapters, we address the three overarching categories—(1) feedback, (2) content, and (3) context—with their corresponding nine teacher actions and thirty-four of the embedded forty-three elements. We do so by providing concrete examples for how to apply a generous representation of the hundreds of instructional strategies expressly for learning and doing science.

Although this text predominantly provides suggestions to support lesson planning around science instruction, we encourage readers to explore the foundational book *The New Art and Science of Teaching* (Marzano, 2017). In doing so, you will likely infuse teachers' content areas and grades with additional strategies. For example, element 16 (highlighting critical information) encompasses the following eleven strategies.

1. Repeating the most important content
2. Asking questions that focus on critical information
3. Using visual activities
4. Using narrative activities
5. Using tone of voice, gestures, and body position
6. Using pause time
7. Identifying critical-input experiences
8. Using explicit instruction to convey critical content
9. Using dramatic instruction to convey critical content
10. Providing advance organizers to cue critical content
11. Using what students already know to cue critical content

Teachers could wisely incorporate all these strategies into various lessons throughout a unit, as they represent sound instructional practice. For example, when teachers continually repeat important information during a lesson and unit, it alerts students to critical content and helps them remember the information. As well, when teachers intentionally and strategically use their tone of voice, gestures, and body position to emphasize salient information, it again highlights what students should remember and focuses their attention on key content. Instead of focusing our attention on these more pervasive strategies—and other such strategies throughout the framework—we provide ideas specific to science. For example, for element 16, we choose the strategy of using visual activities as an opportunity to show how teachers can apply this strategy to teach a science skill, which we detail in chapter 6 (page 77). As readers continue through this text, strategies linked to science take center stage.

About This Book

This book is organized into three parts—(1) feedback, (2) content, and (3) context—mirroring the overarching categories of *The New Art and Science of Teaching* model described earlier in this introduction. The chapters align with the ten teacher actions and then focus on selected elements (from the forty-three) within each action and specific strategies for teaching science.

In part I, chapters 1 and 2 focus on feedback. Chapter 1 pinpoints strategies for providing and communicating clear learning goals, and chapter 2 concentrates on using assessments.

In part II, chapters 3, 4, 5, and 6 focus on content. Chapter 3 looks at conducting direct instruction lessons, chapter 4 on conducting practicing and deepening lessons, and chapter 5 on conducting knowledge application lessons. Chapter 6 focuses on using strategies that appear in all types of lessons.

In part III, chapters 7 and 8 focus on context. Chapter 7 focuses on student engagement, while chapter 8 discusses strategies for organizing the physical layout of the classroom and understanding students' backgrounds and interests. Chapter 9 describes a four-step process for developing teachers' expertise in an effort to increase student learning.

Each chapter includes self-rating scales readers can use to assess their performance on each element addressed in this book. By doing this, they can determine their areas of strength and the areas in which they might want to improve relative to *The New Art and Science of Teaching*. All the self-rating scales in this book have the same format for progression of development. To introduce these scales and help readers understand them, we present the general format of a self-rating scale in figure I.2.

Score	Description
4: Innovating	I adapt strategies and behaviors associated with this element for unique student needs and situations.
3: Applying	I use strategies and behaviors associated with this element without significant errors and monitor their effect on students.
2: Developing	I use strategies and behaviors associated with this element without significant errors but do not monitor their effect on students.
1: Beginning	I use some strategies and behaviors associated with this element but do so with significant errors or omissions.
0: Not Using	I am unaware of the strategies and behaviors associated with this element or know them but don't employ them.

Figure I.2: General format of the self-rating scale.

To understand this scale, it is best to start at the bottom with the Not Using row. Here the teacher is unaware of the strategies that relate to the element or knows them but doesn't employ them. At the Beginning level, the teacher uses strategies that relate to the element but leaves out important parts or makes significant mistakes. At the Developing level, the teacher executes strategies important to the element without significant errors or omissions but does not monitor their effect on students. At the Applying level, the teacher not only executes strategies without significant errors or omissions but also monitors students to ensure they are experiencing the desired effects. We consider the Applying level the level at which one can legitimately expect tangible results in students. Finally, at the Innovating level, the teacher is aware of and makes any adaptations to the strategies for students who require such an arrangement.

Each chapter also contains Guiding Questions for Curriculum Design to help with planning and aid with reflection.

Appendix A provides an overview of *The New Art and Science of Teaching* framework, and appendix B lists the figures and tables featured in this book.

In sum, *The New Art and Science of Teaching Science* is designed to present a science-specific model of instruction within the context of *The New Art and Science of Teaching* framework. We address each of the forty-three elements from the general model within the context of science instruction and provide science-specific strategies and techniques teachers can use to improve their effectiveness and elicit desired mental states and processes from their students.

PART I

Feedback

CHAPTER 1

Providing and Communicating Clear Learning Goals

Feedback is grounded in clarity on what all students will learn and how students will develop proficiency in the learning goals. Students and teachers both engage in ongoing assessment about student proficiency and growth, and clarity is key to effective assessment. Similarly, within *The New Art and Science of Teaching* (Marzano, 2017), teachers should communicate clear learning goals so students understand the progression of knowledge and skills teachers expect them to master and where they are along that progression.

The elements within this first teacher action category of providing and communicating clear learning goals include the following.

- **Element 1:** Providing scales and rubrics
- **Element 2:** Tracking student progress
- **Element 3:** Celebrating success

There are a variety of educational terms used in this realm, including *learning targets*, *objectives*, *rubrics*, *proficiency scales*, and so on. We recommend that districts and schools make choices about the terms they will use, generate working definitions of the terms, and clearly communicate these to teachers, students, parents, and other stakeholders. In this book, we treat these three elements as a coordinated approach to ensure students and teachers can be clear on what students need to learn, the progress on that learning, and how to celebrate the learning.

Element 1: Providing Scales and Rubrics

Scales and rubrics provide the tools for students to understand the progression of knowledge and expectations as the focus for learning. Well-constructed proficiency scales and rubrics provide specific indicators for success. When students and teachers are clear on these indicators, such tools guide meaningful reflection, revision, and action planning for future growth. Proficiency scales can be in various formats. One resource for scales and rubrics that align with the Next Generation Science Standards (NGSS) is *Proficiency Scales for the New Science Standards* (Marzano, Yanoski, & Paynter, 2016). For proficiency scales to be effective in the classroom, teachers must continuously coach students in their application through the review of student work or exemplars.

For element 1 of the model, we select the following specific strategies to address in this chapter. We list additional strategies for element 1 in figure A.1 in the appendix (page 149).

- Clearly articulating and creating scales and rubrics for learning goals
- Using teacher-created targets and scales and implementing routines for using them

It is important to note that simply employing a strategy does not ensure the desired effect on students. Teachers should constantly collect student performance information in order to evaluate the success of any strategy. As indicated in the introduction (page 1), we use the following scale throughout the book for teachers to reflect on the implementation of the element. Teachers may use the self-rating scale in figure 1.1 to rate their current level of effectiveness with the specific strategies for providing scales and rubrics.

Score	Description
4: Innovating	I engage in all behaviors at the Applying level. In addition, I identify those students who do not have an understanding of the proficiency scales or cannot accurately describe their current level of performance and design alternate activities and strategies to meet their specific needs.
3: Applying	I engage in activities to provide students with rubrics and scales without significant errors or omissions and monitor the extent to which students have an understanding of the proficiency scales, and I can accurately describe their current level of performance on the scales.
2: Developing	I engage in activities that provide students with clear rubrics and scales without significant errors or omissions.
1: Beginning	I engage in activities that provide students with clear rubrics and scales but do so with errors or omissions, such as not systematically referring back to the progression of knowledge in the rubric or scale or explaining how daily assignments relate to the learning goal.
0: Not Using	I do not engage in activities that provide students with clear rubrics and scales.

Figure 1.1: Self-rating scale for element 1—Providing scales and rubrics.

Clearly Articulating and Creating Scales or Rubrics for Learning Goals

Rubrics and proficiency scales are used to identify the intended learning for all students, as well as various levels of proficiency. In this chapter, we use the NGSS (NGSS Lead States, 2013) as the basis for these rubrics and scales. While some U.S. states and school districts have not adopted the NGSS, the majority of states have adopted standards that are based on *A Framework for K–12 Science Education* (National Research Council, 2012). This framework was used to create the NGSS, and it outlines three dimensions of science learning as noted in table 1.1. For this reason, we begin with some background context for the arrangement of the standards.

Table 1.1: The Three Dimensions of the Framework

1 Scientific and Engineering Practices
1. Asking questions (for science) and defining problems (for engineering)
2. Developing and using models
3. Planning and carrying out investigations

4. Analyzing and interpreting data
5. Using mathematics and computational thinking
6. Constructing explanations (for science) and designing solutions (for engineering)
7. Engaging in argument from evidence
8. Obtaining, evaluating, and communicating information

2 Crosscutting Concepts
1. Patterns
2. Cause and effect: Mechanism and explanation
3. Scale, proportion, and quantity
4. Systems and system models
5. Energy and matter: Flows, cycles, and conservation
6. Structure and function
7. Stability and change

3 Disciplinary Core Ideas
Physical Sciences
PS1: Matter and its interactions
PS2: Motion and stability—Forces and interactions
PS3: Energy
PS4: Waves and their applications in technologies for information transfer
Life Sciences
LS1: From molecules to organisms—Structures and processes
LS2: Ecosystems—Interactions, energy, and dynamics
LS3: Heredity—Inheritance and variation of traits
LS4: Biological evolution—Unity and diversity
Earth and Space Sciences
ESS1: Earth's place in the universe
ESS2: Earth's systems
ESS3: Earth and human activity
Engineering, Technology, and Applications of Science
ETS1: Engineering design
ETS2: Links among engineering, technology, science, and society

Source: NGSS Lead States, 2013.

From this framework, the NGSS outlines the specific performance expectations for what all students should know and do in science by the completion of a particular grade level or course. Each of these performance expectations includes a Science and Engineering Practice (SEP) that outlines the action that students will do to demonstrate their learning. In addition, the performance expectation includes a Disciplinary Core Idea (DCI) that represents the science concept that will be incorporated. Finally, the Crosscutting Concepts (CCC) represent themes that run through and cross over all science disciplines (Achieve Inc., 2013).

When creating proficiency scales that align with the NGSS and the framework, schools and districts need to decide if these scales will be linked to the performance expectations (PEs) or to the science and engineering practices (SEPs), both of which are outlined in this chapter.

Proficiency scales outline the levels of learning on the three dimensions of the performance expectation. The example in figure 1.2 (page 14) shows a middle school life-science performance expectation (MS-LS1-6; NGSS Lead States, 2013), along with the three dimensions in Table 1.1.

Students who demonstrate understanding can:

MS-LS1-6: Construct a scientific explanation based on evidence for the role of photosynthesis in the cycling of matter and flow of energy into and out of organisms. [Clarification Statement: Emphasis is on tracing movement of matter and flow of energy.] [Assessment Boundary: Assessment does not include the biochemical mechanisms of photosynthesis.]

Science and Engineering Practices	Disciplinary Core Ideas	Crosscutting Concepts
Constructing Explanations and Designing Solutions Constructing explanations and designing solutions in 6–8 builds on K–5 experiences and progresses to include constructing explanations and designing solutions supported by multiple sources of evidence consistent with scientific knowledge, principles, and theories. • Construct a scientific explanation based on valid and reliable evidence obtained from sources (including the students' own experiments) and the assumption that theories and laws that describe the natural world operate today as they did in the past and will continue to do so in the future. **Scientific Knowledge Is Based on Empirical Evidence** • Science knowledge is based upon logical connections between evidence and explanations.	**LS1.C: Organization for Matter and Energy Flow in Organisms** • Plants, algae (including phytoplankton), and many microorganisms use the energy from light to make sugars (food) from carbon dioxide from the atmosphere and water through the process of photosynthesis, which also releases oxygen. These sugars can be used immediately or stored for growth or later use. **PS3.D: Energy in Chemical Processes and Everyday Life** • The chemical reaction by which plants produce complex food molecules (sugars) requires an energy input (i.e., from sunlight) to occur. In this reaction, carbon dioxide and water combine to form carbon-based organic molecules and release oxygen. *(secondary)*	**Energy and Matter** • Within a natural system, the transfer of energy drives the motion and/or cycling of matter.

Source: NGSS, n.d.b.

Figure 1.2: Middle school life science performance expectation.

This proficiency scale shows five levels of learning ranging from 0.0 to 4.0. In addition, intermediate levels are included, although some school districts may choose to use only the whole-number integers. The target or grade-level proficiency is represented by a score of 3.0. This description is exactly as stated in the NGSS and represents the at-grade-level performance we expect all students to eventually attain. The other scores on the scale describe what performance looks like when developing (1.0), approaching (2.0), or moving beyond (4.0) proficiency. For example, a score of 1.0 positively states that, with support and help, a student is able to demonstrate some elements of quality described at the 3.0 level. A score of 2.0 shows the foundational skills, particularly knowledge of science concepts and vocabulary, as well as some early descriptions of how matter is flowing in the system. A score of 4.0 represents an extension of the classroom experience, possibly bringing in other ideas from outside the classroom or making connections with material learned at other times in the school year. Teachers and students can use this clear picture of intended outcomes to guide what will be taught and, more importantly, to see what learning students must do to reach proficiency.

In this performance expectation example, students are asked to construct an explanation (SEP 6) in which they include several aspects of Organization for Matter and Energy Flow in Organisms (LS1.C) as well as the crosscutting concept of Energy and Matter. In addition, a related DCI (PS3.D) can be incorporated in the ensuing instruction and learning. At the middle school level, a proficiency scale could be designed as shown in figure 1.3.

Score	Middle School—MS-LS1-6
4.0	The student will make connections between the terminology from the current unit of study, as well as demonstrate a clear connection with the process of cellular respiration, indicating the connections between reactants and products. In addition, the student can discuss the results when photosynthesis can't keep pace with the release of carbon dioxide into the atmosphere, leading to scenarios such as the greenhouse effect and global climate change.
3.5	In addition to score 3.0 performance, the student has partial success at score 4.0 content.
3.0	The student will: MS-LS1-6 • Construct a scientific explanation based on evidence for the role of photosynthesis in the cycling of matter and flow of energy into and out of organisms
2.5	The student has no major errors or omissions regarding score 2.0 content and partial success at score 3.0 content.
2.0	The student will: MS-LS1-6 • Recognize or recall specific vocabulary from instruction (for example: plant, oxygen, carbon dioxide, sugars, light energy, flow, leaf, photosynthesis, cycle) • Describe how matter cycles between plants and other organisms through the process of photosynthesis
1.5	The student has partial success at score 2.0 content and major errors or omissions regarding score 3.0 content.
1.0	With help, the student has partial success at score 2.0 content and score 3.0 content.
0.5	With help, the student has partial success at score 2.0 content, but not at score 3.0 content.
0.0	Even with help, the student has no success.

Source for standard: NGSS, n.d.b.

Figure 1.3: Middle school science proficiency scale.

For additional explanations of the architecture of the standards and for more proficiency scales, see *Proficiency Scales for the New Science Standards* (Marzano, Yanoski, & Paynter, 2016). In addition, the NGSS has provided evidence statements for the performance expectations that you can access online (www.nextgenscience.org/evidence-statements; NGSS, n.d.a).

Using Teacher-Created Targets and Scales and Implementing Routines for Using Them

Many teachers and school districts choose to construct targets and scales that are more overarching and connect directly to the science and engineering practices (SEPs). In this way, the targets, scales, and routines can transcend a particular unit of study or even grade level. Teachers can then use these targets and scales to demonstrate the recursive nature of the SEPs and CCCs and how they are used to link the science concepts together over the course of a student's learning experience. For example, using the same general learning outcome noted previously, a teacher could focus on proficiency levels of the science and engineering practice of constructing scientific explanations, as shown in figure 1.4.

Exceeding (4)	Meeting (3)	Approaching (2)	Developing (1)
I can construct an explanation for a scientific phenomenon using all success criteria in unfamiliar contexts or making connections to related science concepts.	I can construct an explanation for a scientific phenomenon using all success criteria in familiar contexts.	I can construct an explanation for a scientific phenomenon using some success criteria in familiar contexts.	I can construct an explanation for a scientific phenomenon in familiar contexts with support.

Success Criteria:

- Create an accurate claim.
- Use multiple valid and reliable sources of evidence.
- Provide clear and complete reasoning that supports the claim and that clearly connects with science concepts and vocabulary (these can be listed out or not; for example, *plant, oxygen, carbon dioxide, sugars, light energy, flow, leaf, photosynthesis, cycle*).
- Clearly demonstrate a connection to the crosscutting concept.

Figure 1.4: Proficiency scale and success criteria for constructing scientific explanations.

In this approach, these same teacher-generated scales and success criteria can be used throughout the course and students' science experience. Students are then able to reflect on their learning and provide evidence for how their performance matches (or does not match) the proficiency expectations. When asked to thoughtfully review and reflect on their work, students are often able to see where they have not fully met all success criteria (Reibel & Twadell, 2019). For example, following with the same example here for constructing scientific explanations, a student could engage in reflective practices or engage in conversations with peers or the teacher using the format outlined in the section on element 22 (Peer Response Groups, page 95).

For additional success criteria ideas, as well as what to look for in student work at all grade levels, refer to the NGSS *Appendix F: Science and Engineering Practices in the NGSS* (NGSS, 2013).

Element 2: Tracking Student Progress

Tracking student progress in the science classroom is similar to tracking student progress in any content area: the student receives a score based on a proficiency scale, and the teacher uses the student's pattern of scores to "provide each student with a clear sense of where he or she started relative to a topic and where he or she is currently" (Marzano, 2017, p. 14). For each topic at each applicable grade level, teachers should construct a proficiency scale (or learning progression). Such a scale allows teachers to pinpoint where a student falls on a continuum of knowledge and skill, using information from assessments. Figure 1.5 presents the self-rating scale for element 2 (tracking student progress).

Score	Description
4: Innovating	I engage in all behaviors at the Applying level. In addition, I identify those students who are not aware of what they must do to improve and design alternate activities and strategies to meet their specific needs.
3: Applying	I engage in activities to track student progress without significant errors or omissions and monitor the extent to which students are aware of what they must do to improve their current status.
2: Developing	I engage in activities to track student progress without significant errors or omissions.
1: Beginning	I engage in activities to track student progress but do so with errors or omissions, such as not keeping track of the progress of individual students and not making students aware of their individual progress.
0: Not Using	I do not engage in activities to track student progress.

Figure 1.5: Self-rating scale for element 2—Tracking student progress.

This section illustrates the following concrete examples for science instruction associated with strategies about tracking progress. (For all the strategies related to this element, see appendix A, page 149.)

- Designing assessments that generate formative scores
- Charting student progress and class progress

Designing Assessments That Generate Formative Scores

Frequent formative assessments are crucial in providing valuable learning information to both teachers and students. In *Embedded Formative Assessment*, Dylan Wiliam (2018) suggests that "an assessment functions formatively to the extent that evidence about student achievement is elicited, interpreted, and used by teachers, learners, or their peers to make decisions about the next steps in instruction that are likely to be better, or better founded, than the decision they would have made in the absence of evidence" (p. 43). In order to view and evaluate evidence, all stakeholders in the process need a defined common approach that establishes performance expectations, lays out indicators of success, and provides a way to evaluate quality with calibrated proficiency scales. In any successful formative feedback loop, teachers and students gain information to inform a change in practice. For students, this may involve a new study approach. For teachers, the information may lead to re-teaching a skill or returning to the skill in a new context.

Proficiency scales, as noted in the previous sections, allow teachers and students to monitor learning and growth over time. Formative assessments must be linked directly to the performance expectation, science and engineering practice, or disciplinary core idea in such a way that a teacher can create an assessment related to the level or depth of knowledge at which the student is showing proficiency. In subsequent chapters, we highlight assessment ideas, such as in element 4 (Using Informal Assessments of the Whole Class, page 23), element 5 (Using Formal Assessments of Individual Students, page 27), and element 19 (Reflecting on Learning, page 88) to illustrate the strategies and assessments that can be used. As readers move through this book, they should be looking for and considering opportunities to assess student learning and provide formative feedback to and with students.

Charting Student Progress and Class Progress

It is imperative that both students and teachers are clear on the proficiency level at which students are performing and how they are growing over time. This monitoring can apply to specific performance expectations or growth in recursive skills such as science and engineering practices. Figure 1.6 shows the progress monitoring both teachers and students could do for an individual performance expectation. Each student should monitor his or her own score, which could have been generated by individual self-reflection or from teacher feedback. In addition, the teacher can track the entire class performance to look for patterns and trends, and plan interventions.

Performance Expectation: 3-PS2-2—Make observations and/or measurements of an object's motion to provide evidence that a pattern can be used to predict future motion. [Clarification statement: Examples of motion with a predictable pattern could include a child swinging in a swing, a ball rolling back and forth in a bowl, and two children on a see-saw.] [Assessment boundary: Assessment does not include technical terms such as period and frequency.]				
Student	**Assessment 1**	**Assessment 2**	**Assessment 3**	**Assessment 4**
Billy	Approaching (2)	Meeting (3)	Approaching (2)	Approaching (2)
Trey	Meeting (3)	Meeting (3)	Meeting (3)	Meeting (3)
Unnathi	Approaching (2)	Meeting (3)	Approaching (2)	Meeting (3)
Wen	Meeting (3)	Meeting (3)	Approaching (2)	Approaching (2)
Julia	Approaching (2)	Meeting (3)	Approaching (2)	Meeting (3)

Source for standard: NGSS, n.d.b.

Figure 1.6: Student progress over time.

When reviewing scores in this way, the teacher is able to gain insights into student performance by seeing some patterns in the evidence. For example, Trey has demonstrated clear proficiency with each assessment. This could be an opportunity to enrich his learning, such as by asking him to make connections to other learning throughout the year, include more technical terms such as frequency and period in his work, or describe scenarios that haven't been considered in class. For other students, such as Unnathi, Billy, and Julia, the teacher and student would want to have additional information about why the proficiency scores seem to be moving up and down. This could be due to a need for a deeper understanding of science vocabulary, clarity on the connections to the crosscutting concept, or the accuracy and confidence in making observations or measurements. On the other hand, Wen's assessment data show a somewhat concerning pattern, possibly indicating that he is losing interest or motivation. He demonstrated proficiency early in the learning but has tapered off to below proficient in his most recent assessments.

In addition to providing insights on individual students and their learning needs, tracking student performance in this way provides insights on the assessments themselves, with a clear pattern of proficiency by all students on assessment 2, and only one student meeting proficiency on assessment 3. This could present insights on the depth of knowledge required on these assessments or, in some cases, the validity of the assessment itself. If these are common assessments used with other teachers in the school, it is helpful to see if similar patterns are observed in those classrooms, making this a learning experience for students and teachers alike.

Element 3: Celebrating Success

Once a strong system for tracking student progress is in place, teachers and students have a great deal of rich information with which to celebrate success. In the science classroom, celebrations should focus on student proficiency and growth. These celebrations should be authentic, but the teacher should be mindful of recognizing *all* students throughout the learning process. Too often, classroom "superstars" get recognized and many students don't get the encouragement they need to persist through difficult work.

In science classrooms (where it seems teachers are often White males), it is especially important for teachers to examine their own biases and be culturally responsive. In addition, future representation in STEM (science, technology, engineering, mathematics) fields could depend on building self-efficacy and grit by shining a light on student strengths. Currently, Black and Hispanic people are severely underrepresented in STEM fields, and women are underrepresented in all fields but healthcare (Fry, Kennedy, & Funk, 2021). Recognizing moments when students "get it" (proficiency) and when they improve can build student pride in their scientific work. Figure 1.7 presents the self-rating scale for this element.

Score	Description
4: Innovating	I engage in all behaviors at the Applying level. In addition, I identify those students who do not exhibit a sense of pride in their accomplishments and design alternate activities and strategies to meet their specific needs.
3: Applying	I engage in activities to celebrate students' success without significant errors or omissions and monitor the extent to which students have a sense of pride in their accomplishments.
2: Developing	I engage in activities to celebrate students' success without significant errors or omissions.
1: Beginning	I engage in activities to celebrate students' success but do so with errors or omissions, such as acknowledging students' status but not growth and not providing continual verbal encouragement.
0: Not Using	I do not engage in activities to celebrate students' success.

Figure 1.7: Self-rating scale for element 3—Celebrating success.

In this section, we feature the following strategies for celebrating success.

- Knowledge gain celebration
- Verbal feedback

Knowledge Gain Celebration

Celebrations of knowledge gains should focus on improvement related to the proficiency standards set for student learning (see element 1 in this chapter, page 11). Specifically, these strategies celebrate gains in science knowledge and skills. Proficiency scales provide tangible and visible measures to chart these gains, and it's important for teachers and students to celebrate this learning. As outlined in the proficiency scales earlier in this chapter, this could include moving from a level of 2.0, where support is needed (only some of the success criteria are met), to a level of 3.0, where the full range of success criteria are consistently met. Further, students can celebrate when they hit a level of 4.0, making extensions and connections beyond the current unit of study or linking the current ideas to others in the course or even to their lives outside the classroom. Using proficiency scales to celebrate these knowledge gains provides a measurable, repeatable, and objective basis for celebration.

Verbal Feedback

Feedback, when given effectively, can have a dramatic impact on student learning. In his often-cited research compilation *Visible Learning*, John Hattie (2009) identifies feedback as a significant factor in learning, with an effect size of 0.73, well above the "hinge point" of 0.40, which is deemed as the average effect size of various education practices that impact learning (Hattie, 2009). Further, John Hattie and Helen Timperley (2007) highlight the value of feedback for student learning gains. In particular, high-information feedback provides a specific path forward toward learning goals and contains clear information about how work compares to the learning goals (or success criteria). The authors also note the importance of timing feedback with regard to where students are in the learning cycle. Verbal feedback provides the science teacher with the opportunity to continually monitor student learning and continually guide with questions, suggestions, and corrections.

One benefit of monitoring student learning over time, as previously referenced in element 2, is that it allows students to celebrate not only educational attainment but learning gains as well. Teachers can do this at the end of a lesson or unit of study, but it is often most powerful when done in the midst of learning. It is important that teachers and students recognize the value of persistence and hard work, and how those link to student success. For example, when students run into a challenge with a laboratory investigation, it is powerful for the teacher to praise students for persisting in a systematic way to move through the challenge. In her often-cited work, Carol Dweck (2016) describes how praising student effort and hard work are connected to promoting a *growth mindset*. In contrast, praising students as "smart" or saying that learning comes easily promotes a *fixed mindset*, often resulting in a lack of motivation or confidence.

When praising students for their persistence, teachers should go beyond surface-level praise. It is most beneficial when this feedback is descriptive and linked to a measurable indicator of persistence. For example, the teacher can praise the number of times a student repeats a task until reaching proficiency or naming the specific actions observed, which students can replicate in future work. Further, students can reflect on how they have increased in proficiency in science and engineering practices (see the teacher-generated scale in figure 1.4, page 16, for constructing scientific explanations).

In addition to teacher feedback, it is powerful for students to reflect on their progress and recognize the gains they have made. Teachers should reinforce the need for students to continually reflect on their own learning and articulate how their current state of proficiency matches the standard their teachers set. This metacognitive process helps make students' thinking visible to both themselves and to the teacher. Jan Chappuis (2015) encourages teachers to have students ask, "Where am I going? Where am I now? How can I close the gap?"

This is particularly powerful when monitoring growth in a science practice such as engaging in argument from evidence. By providing students an opportunity to review their previous writing samples, identify growth in specific success criteria, and note where they have made significant gains, students can see their own gains in learning. Similarly, peers can provide feedback and comment on the progress they see in their peers' work, especially when peer assessment is built into the class as a regular part of instruction, assessment, and feedback.

Finally, self and peer feedback and reflection can promote social-emotional learning (SEL), as outlined in the five CASEL competencies and as described in the CASEL framework.

1. *Self-awareness:* The abilities to understand one's own emotions, thoughts, and values and how they influence behavior across contexts.

2. *Self-management:* The abilities to manage one's own emotions, thoughts, and behaviors effectively in different situations and to achieve goals and aspirations.

3. *Social awareness:* The abilities to understand the perspectives of and empathize with others, including those from diverse backgrounds, cultures, and contexts.

4. *Relationship skills:* The abilities to establish and maintain healthy and supportive relationships and to effectively navigate settings with diverse individuals and groups.

5. *Responsible decision making:* The abilities to make caring and constructive choices about personal behavior and social interactions across diverse situations. (Collaborative for Academic, Social and Emotional Learning, n.d.).

These are skills that most teachers and parents hope to develop in students, and building in times for reflection and celebration promote these skills. In addition, there are clear learning gains when teachers focus effort on enhancing students' social-emotional learning competencies (Durlak, Weissberg, Dymnicki, Taylor, & Schellinger, 2011). In this particular study involving over 270,000 students, the authors noted an 11 percentile increase in achievement when students were engaged in a school-based social and emotional learning program (Durlak et al., 2011).

GUIDING QUESTIONS FOR CURRICULUM DESIGN

When teachers engage in curriculum design, they consider this overarching question for communicating clear goals and objectives: *How will I communicate clear learning goals that help students understand the progression of knowledge I expect them to master and where they are along that progression?* Consider the following questions aligned to the elements in this chapter to guide your planning.

- **Element 1:** How will I design scales and rubrics?

  _____

  _____

  _____

  _____

  _____

  _____

  _____

- **Element 2:** How will I track student progress?

  _____

  _____

  _____

  _____

  _____

  _____

  _____

- **Element 3:** How will I celebrate success?

_____

_____

_____

_____

_____

_____

_____

Summary

Effective feedback—the first of three overarching categories in part 1—begins with clearly defined and articulated learning goals. When teachers make expectations transparent so students understand what they are to learn within a lesson or unit, students can determine how well they are performing and what they need to do to improve. Once teachers focus on providing and communicating clear learning goals, they can direct their attention to using effective assessments.

CHAPTER 2

Using Assessments

During science instruction, some teachers use assessments only as evaluation tools to quantify students' current status relative to mastery of specific science concepts and skills. While this is certainly a legitimate use of assessments, their primary purpose should be to provide students with feedback they can use to improve. When teachers use assessments to their full capacity, students understand how their assessment results and grades relate to their status on specific progressions of knowledge and skill they are expected to master.

There are two elements within this category.

- **Element 4:** Using informal assessments of the whole class
- **Element 5:** Using formal assessments of individual students

Element 4: Using Informal Assessments of the Whole Class

Rather than formal assessments of individual students—the emphasis of element 5—the focus here is on informally assessing the whole class. Informal assessments occur when the teacher takes a holistic view of students' progress throughout the classroom. While this can be done in many ways, some of which are outlined in the following section, the intent is to inform instruction and pacing rather than provide a score or grade. As students engage in science, whether it be conducting a lab investigation, constructing explanations with peers, or analyzing scientific data, the teacher is continuously monitoring students' progress, asking questions, and observing students as they work. The teacher prompts and guides the learning with specific feedback based on what he or she sees in student work or responses. This provides the teacher with a barometer of how students are progressing with specific skills along a continuum of growth to inform their instructional moves. Figure 2.1 (page 24) presents the self-rating scale for this element so teachers can gauge their professional performance.

Within this element, we show teachers how to apply the following strategies in the classroom.

- Confidence rating techniques
- Response boards

As a reminder, refer to figure A.1 in the appendix (page 149) for a complete listing of all strategies related to each of the forty-three elements.

Score	Description
4: Innovating	I engage in all behaviors at the Applying level. In addition, I identify those students who are not using the whole-class feedback to set personal and group goals and design alternate activities and strategies to meet their specific needs.
3: Applying	I engage in activities to informally assess students without significant errors or omissions and monitor the extent to which students use the feedback to set personal and group goals.
2: Developing	I engage in activities to informally assess the class as a whole without significant errors or omissions.
1: Beginning	I engage in activities to informally assess the class as a whole but do so with errors or omissions, such as not focusing on content important to students' learning or not providing feedback regarding the class's status on specific progressions of knowledge.
0: Not Using	I do not engage in activities to informally assess the class as a whole.

Figure 2.1: Self-rating scale for element 4—Using informal assessments of the whole class.

Confidence Rating Techniques

A key reason to use informal assessments is to give students repeated opportunities to compare their proficiency level against the established and communicated success criteria. This is particularly important in classrooms where students are asked to demonstrate proficiency in science-specific skills, such as the science and engineering practices. Students are able to share how confident they are in independently performing the skill or practice within the given science context. This confidence rating is not as effective if the science learning is focused on rote memorization or factual recall; in those situations, a student can either recall that information or they can't. Rather, confidence ratings are most effective when students understand the proficiency scale (or range) that relates to their performance. There are several ways to provide students with opportunities to reflect on their confidence level in performing the science skill, and teachers should routinely use these tools throughout a lesson and unit of study.

A quick, simple, and effective whole-class monitoring strategy is *fist to five*. With this strategy, the teacher asks students to rate their confidence on a concept or skill by holding up the number of fingers that represent their confidence. Holding up five fingers means complete confidence and readiness to move on. Conversely, a fist indicates the student's need to stop and receive some form of teacher or collaborative group intervention, reteaching, practice, or clarification.

One of the simplest and most straightforward (yet effective) tools for students is to reflect on their confidence to demonstrate proficiency on a target, whether that be knowledge-based or skills-based learning targets. The next two items could be done with the whole class but could also begin with individual students reflecting before reporting out in the larger group. For example, students can individually respond to and reflect on prompts, such as those in figure 2.2, to analyze their progress and confidence levels.

Learning Target	Confidence Level				Next Steps in My Learning Toward Mastery
Knowledge Target Samples					
I can state the properties of solids, liquids, and gases.	1	2	3	4	
I can state the relationship between mass and volume when determining the density of a solid or liquid.	1	2	3	4	
I can match cellular organelles to the function they carry out in a cell.	1	2	3	4	
Skill Target Samples					
I can construct an effective scientific explanation by making an accurate claim, identifying relevant evidence, and providing thorough reasoning supported by the science concepts we have learned.	1	2	3	4	
I can construct graphical representations of lab data by choosing the correct type of graph, scaling the axes appropriately, plotting the data, and labeling all parts of the graph.	1	2	3	4	
I can plan a scientific investigation by identifying the relevant variables, creating a procedure that tests a relationship, and describing the necessary equipment and procedures to conduct the experiment.	1	2	3	4	
1 = I need a great deal of support to master this. 2 = I've almost got this! 3 = I've got this; I am very confident. 4 = I could teach this to others and apply this skill to new contexts.					

Figure 2.2: Confidence levels progress chart.

Visit go.SolutionTree.com/instruction for a free blank reproducible version of this figure.

A final option for confidence-rating techniques is specific practice sets. For example, students can complete a given set of practice problems or questions. It is important to note that this is not simply "drill and kill," in which students are repeating a task until the point of overlearning. Rather, the focus here is on practice that is done in class, focuses on skill development, and involves mindful reflection by students as they develop proficiency in the skill (Lang, 2021). After completing the questions, students rate themselves on a simple scale (similar to that in figure 2.2). In cases in which students have not yet mastered or are not yet confident, they state the steps they will complete to develop more knowledge or a skill before moving forward. Use the tools and strategies noted in the next section to help students report out these confidence ratings to the teacher. This process of reflection both guides the students in mindful learning and provides insights and evidence for teachers to make informed decisions about future classroom segments or tasks students complete.

Response Boards

Teachers can use a variety of strategies to quickly gather evidence of learning from the entire class. Sarah A. Nagro, Sara D. Hooks, Dawn W. Fraser, and Kyena E. Cornelius (2018) find the value in using these whole-class strategies, particularly in an inclusive classroom, to be an effective way to engage all students in active learning and curb unwanted behaviors. Using response boards is one such strategy for science teachers to get a snapshot of what students know throughout the classroom. This is particularly helpful when students are learning and practicing key concepts related to the science disciplinary core ideas. These concepts are later expressed and assessed more formally in conjunction with the science practices, but this work can serve as an important preparatory piece. In addition to informing the teacher, students are able to retrieve key information, practice important skills, and reflect on their mastery of learning.

In this strategy, students use small whiteboards or an equivalent tool to write a response to questions posed either by the teacher or by other students. Each student (or pair of students) generates a response to the question and holds up the whiteboard for the teacher to see. The teacher can choose to have students hold up the boards for an extended time, allowing other students to see the range of answers in the room. Alternatively, the teacher can ask for a very fast show, allowing for a quick scan of the room to maintain student privacy. With technology tools, such as tablet computers (for example, iPads), students also could write their responses on the screens and hold them up for the teacher or peers to see. Whether using response boards or digital devices, this strategy has the advantage over a choral response in that the teacher is able to quickly gather assessment evidence for each student.

Another effective and somewhat related use of response boards is to have students map out their thinking on practice problems, such as those assigned for homework. For example, in the section Confidence Rating Techniques (page 24), we suggest having students rate their confidence on stating the relationship between mass, volume, and density. Students might use response boards to calculate mass, volume, or density when given the other two variables. Many teachers use the density triangle shown in figure 2.3 to support students as they see these relationships.

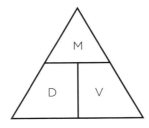

Figure 2.3: Density triangle.

To find the unknown variable, students cover the variable they are looking to solve, which then gives the relationship or equation needed to solve for the unknown. For example, if given density and volume, students solve for mass by covering the M and noting that density is multiplied by volume. As students progress, they also see how the units cancel out when doing these calculations.

Teachers can assign individual students, pairs, or small groups of students to work out the problem and highlight key steps. Students then present to their peers how they completed the work. The most important and beneficial aspect of this strategy is having students explain their thinking and decision-making process when moving through each problem. Again, this is a highly effective way for teachers and students to identify gaps in learning and then call on the collective expertise of other students in the class to provide insights and feedback.

With the widespread use of technology tools, students can also use a variety of polling devices or other interactive tools to respond to questions. These have the benefit of built-in novelty, competition, or maintaining a record of student responses. Teachers should be careful to select the best technology tool for the activity, considering the types of responses that will most effectively produce evidence of student understanding. Currently available tools include Pear Deck (peardeck.com), Quizizz (quizizz.com), Socrative (socrative.com) Formative (formative.com; community.formative.com), Classkick (classkick.com), Google Classroom (classroom.google.com), and ShowMe (showme.com).

Element 5: Using Formal Assessments of Individual Students

To individually show evidence that students can demonstrate their understanding of an item or a set of related items on a proficiency scale, the teacher issues formal assessments. While these assessments also inform the next steps in teaching and learning for students and teachers (as do informal assessments discussed in element 4, page 23), formal assessments often lead to a score or preview an assessment that teachers will use for an evaluative score. These more formal assessments typically take place further along the learning continuum and integrate the prior learning and skills to that point. Formal assessments gather evidence that leads to feedback and tracking progress for individual students.

In the context of science, it's important that these assessments represent the three-dimensional nature of science learning represented in the NGSS. As such, teachers should gather evidence about student proficiency related to disciplinary core ideas, science practices, and crosscutting concepts. Figure 2.4 shows the self-rating scale teachers can use for this element.

Score	Description
4: Innovating	I engage in all behaviors at the Applying level. In addition, I identify those students who are not independently using their scores on these assessments as information with which to improve their learning and design alternate activities and strategies to meet their specific needs.
3: Applying	I engage in activities to formally assess individual students without significant errors or omissions and monitor the extent to which students are independently using their scores on these assessments as information with which to improve their learning.
2: Developing	I engage in activities to formally assess individual students without significant errors or omissions.
1: Beginning	I engage in activities to formally assess individual students but do so with errors or omissions, such as not using the information gained by these assessments to provide individual students with helpful feedback regarding how to improve.
0: Not Using	I do not engage in activities to formally assess individual students.

Figure 2.4: Self-rating scale for element 5—Using formal assessments of individual students.

To address this element, teachers might use the following strategies from the seven aligned to this element, as detailed in appendix A (page 149).

- Common assessments designed using proficiency scales
- Assessments involving selected-response or short constructed-response items
- Observations of students

Common Assessments Designed Using Proficiency Scales

As outlined in elements 1 and 2 in chapter 1 (page 11), well-constructed proficiency scales guide the learning process for both students and teachers. We have found it effective to include the proficiency scales directly on the assessment itself. Students are then able to better understand what they will be assessed on. This also promotes feedback and self-reflection directly related to the proficiency scale. For example, when exploring chemical changes and physical changes, students are asked to determine whether a new substance was formed. Fifth-grade teachers, for example, can create common assessments that allow students to demonstrate their proficiency level. Students also are able to see the expectations for working at a 2.0 or 3.0 level based on the proficiency scale in figure 2.5. In this example, students explore whether mixing two substances results in a new substance as well as make observations during the process. Figure 2.5 shows the proficiency scale for this task.

Grade 5	
Score 3.0	The student will: • 5-PS1-4—Conduct an investigation to determine whether the mixing of two or more substances results in new substances (for example, observe the mixing of two or more substances and decide whether a chemical reaction has occurred).
Score 2.0	The student will: • Recognize or recall specific vocabulary (for example, *chemical reaction, mix, substance*). • Describe the signs or signals that indicate a chemical reaction.

Source: Marzano et al., 2016, p. 79.
Source for standard: NGSS Lead States, 2013.

Figure 2.5: Example proficiency scale.

In this assessment, students need to predict whether mixing two substances will create a new substance (chemical reaction) or if it will simply be a mixture. To achieve a score of 2.0, students need to use words and terms such as *chemical reaction, mix,* and *mixture* and then describe the signs that indicate a chemical reaction occurred. To achieve a 3.0 score, students need to be able to independently conduct the investigation, generate lab data, and then explain their conclusion using the data they generated. This leads them to correctly identify what took place in the experiment, most notably if a mixture was formed or if a chemical reaction resulted in the formation of a new substance. Some common substances that can be mixed by fifth-grade students for this investigation include water and lemonade powder, baking soda and vinegar, water and flour, and vinegar and olive oil.

When a team of teachers uses a common assessment such as this, they identify common misconceptions demonstrated in the student work. In addition, these teachers are able to develop a collective response based on the pattern they see in student work. For example, students who continue to struggle with basic terminology might benefit from additional practice and instruction on vocabulary, whereas other students might use the vocabulary correctly but have trouble identifying the indicators of a chemical reaction. The teachers then plan the next steps in learning, which could include different teachers taking various groups of students for reteaching and additional practice.

Assessments Involving Selected-Response or Short Constructed-Response Items

Teachers most often make use of selected-response and short-constructed response items on formal paper-and-pencil assessments. Selected-response items include formats such as multiple choice, true or false, matching, and so on. The benefit of these assessments is that they can prove insights on a large amount of content in a relatively short period of time. There are some negative aspects of using these response items, such as the challenge in surfacing the underlying thinking or misunderstandings that brought students to choose a particular item (or even that students simply guessed correctly). In addition, high-quality items are challenging to write. Constructed-response items include short written or verbal explanations, creating a simple drawing, or making a graph. These tasks provide additional insights into students' thinking and more readily incorporate science and engineering practices, or other complex skills. One challenge is the time needed to score them along with the need to be consistent between students and between teachers. (Visit the Educational Testing Service at www.ets.org/Media/Research/pdf/RD_Connections11.pdf for more guidance on using constructed-response test questions.) In addition to the more traditional uses of selected-response and constructed-response items, science teachers can effectively use these items for students to respond to an observation they make and then respond in the moment.

For example, science teachers might begin class with a demonstration of a scientific phenomenon students learned during a previous lesson that included a recurring theme of making keen observations. Following the demonstration, students practice by including details they observed about the phenomenon and explaining the scientific reasoning behind their explanation. In addition, teachers generate multiple-choice questions that hone in on particular misconceptions students have about the scientific phenomenon and then engaging students in rich conversations that explore their misconceptions and come to a more accurate understanding.

An excellent resource for these types of assessments and questions can be found in the series of formative assessment probe books by Page Keeley (NSTA Press). These include titles such as *Uncovering Student Ideas in Physical Science: 32 New Matter and Energy Formative Assessment Probes, Volume 3* (Keeley & Cooper, 2019), *Uncovering Student Ideas in Astronomy: 45 New Formative Assessment Probes* (Keeley & Sneider, 2012), and *Uncovering Ideas in Life Science: 25 New Formative Assessment Probes, Volume 1* (Keeley, 2011). There are numerous books in the series and several volumes across science disciplines. These books span all grade levels and science disciplines (Keeley, n.d.; www.nsta.org/book-series/uncovering-student-ideas-science).

Observations of Students

Since doing science is highly dependent on particular skills, it is important for students to practice science-specific skills and for teachers (and peers) to observe students carrying out these skills. Such skills include actions like measuring accurately (with a ruler, graduated cylinder, and so on), performing a titration, using an electronic balance, focusing a microscope, using dissection tools, or constructing an electric circuit.

Teachers explain and demonstrate how to perform each skill and then observe students as they practice and refine these skills. When students are engaged in science, teachers or peers can assess progress through observation and conversation. In his book *Rebooting Assessment*, Damian Cooper (2022) provides guidance on prompts and look-fors while students are demonstrating complex skills and being assessed during conversations and performances. These skills could include focusing a microscope, using lab equipment, or gathering data. The teacher can provide success criteria to focus the assessment, feedback, and coaching to the most important elements of the skill. This is particularly helpful when peers do the observations.

It is important that both teachers and students are specific in their observations, data collection, and feedback about the particular lab skill. This feedback should focus on agreed-on success criteria for that skill. With specific observations, possibly with the aid of short videos taken on a phone or other device, students receive specific feedback on how they are growing in the skill. A key component of this feedback comes from developing rubrics and success criteria that describe what good work is on that skill and then providing feedback on how the students might improve. This process forces teachers to be specific about identifying the success criteria that allow students to master the skill.

One strategy for this observation is for the teacher (or a student peer) to make a half-sheet of paper that contains the proficiency scale (rubric). As the teacher moves throughout the room, he or she leaves feedback for each student, including any descriptive feedback that outlines a skill gap or highlights an aspect of skill improvement. It could also include giving a score value of 1–4 and then having students reflect on the aspects of their work that led to that score. For example, if the teacher gives a score of 2, students could use their paper to reflect on how their work compares to the level-2 criteria. Alternatively, the teacher could leave only descriptive feedback and then ask students to compare this feedback and their own reflections to the proficiency scale, allowing them to both assign a score to their work and to make plans for improving on the skill.

GUIDING QUESTIONS FOR CURRICULUM DESIGN

This overarching question can guide teachers when using assessments: *How will I design and administer assessments that help students understand how their test scores and grades are related to their status on the progression of knowledge I expect them to master?* Consider the following questions aligned to the elements in this chapter to guide your planning.

- **Element 4:** How will I *informally* assess the whole class?

_____

_____

_____

- **Element 5:** How will I *formally* assess individual students?

_____

_____

_____

Summary

Assessments are feedback tools for both students and teachers. Used well, they serve as instructional and evaluation mechanisms by offering students information on how to advance their understanding of science concepts and skills, giving teachers a vehicle for assisting students to do so. By informally assessing the whole class and formally assessing individual students in various ways, teachers can support students in this learning progression. To gain proficiency in science concepts and skills, teachers design and deliver direct instruction lessons, as explained in the next chapter.

PART II
Content

CHAPTER 3

Conducting Direct Instruction Lessons

The third design area in *The New Art and Science of Teaching* framework involves addressing content in the science classroom through direct instruction lessons. This type of instruction commonly suffers from the perception that it is a teacher-directed presentation in lecture format, but this is far from the truth. As this chapter illustrates, direct instruction has a number of essential components that science teachers can deliver using a wide variety of strategies.

Regardless of the specific strategies teachers use, the net effect of direct instruction should be that students understand the key parts of new science ideas and how, together, they form a unified theory or explanation. Inquiry-based learning definitely has its place, but direct instruction prior to this method may be important for subsets of learners that include ELs (English learners). Research shows that ELs do not show the same achievement levels as non-ELs in inquiry-based lessons, and some direct instruction methods prior to inquiry-based activities may be of great benefit (Estrella, Au, Jaeggi, & Collins, 2018). In contrast to inquiry-based instruction, direct instruction is teacher organized and teacher supported.

In the science classroom, this often involves engaging students in an activity in which they must organize newly acquired vocabulary into a coherent written explanation. A teacher also may ask students to apply newly acquired content to a variety of novel scenarios to develop critical thinking skills. Through these direct instruction activities, students develop the capacity to construct their own knowledge and transfer it to the appropriate contexts.

In this chapter, we address teaching individual skills in conjunction with metacognitive reflection in order to piece together skills to develop a fundamental understanding of complex scientific concepts or principles. We highlight the importance of this design area to the scientific learning process, as it is foundational for deeper learning and future student-to-student collaborations. This category includes the following elements.

- **Element 6:** Chunking content
- **Element 7:** Processing content
- **Element 8:** Recording and representing content

Element 6: Chunking Content

Chunking content is an historically successful approach to various scientific areas of study (Marzano, 2017). When chunking, learners break down larger concepts into smaller, more manageable pieces that they can reassemble or reorganize later. Smaller chunks are easier to hold in working memory, and each chunk can be more easily characterized with street language or visualizations. For example, the complex biological process of photosynthesis can be broken down memorably into *solar panels* (pigments), *battery charging* (conversion of NADP+ to NADPH and ADP to ATP), and the *sugar factory* (Calvin cycle). These three distinct chunks can be further broken down and assigned student-friendly language or visual cues.

Learning new information can be overwhelming; however, breaking it down into manageable increments can facilitate student learning. Figure 3.1 presents the self-rating scale for this element.

Score	Description
4: Innovating	I engage in all behaviors at the Applying level. In addition, I identify those students for whom the chunking process is not helping them understand the content and design alternate activities and strategies to meet their specific needs.
3: Applying	I engage in activities to chunk content when presenting new information without significant errors or omissions and monitor the extent to which the chunking process is helping students understand the new content.
2: Developing	I engage in activities to chunk content when presenting new information without significant errors or omissions.
1: Beginning	I engage in activities to chunk content when presenting new information but do so with errors or omissions, such as not breaking the new content into small enough chunks that students can easily process or breaking content into chunks that are too small.
0: Not Using	I do not engage in activities to chunk new content when presenting new information.

Figure 3.1: Self-rating scale for element 6—Chunking content.

In this section, we specifically highlight these strategies related to chunking science content.

- Using preassessment data to plan for chunks
- Presenting content in small, sequentially-related sets
- Allowing for processing time between chunks

Using Preassessment Data to Plan for Chunks

When presenting new information and concepts to students, it is important to consider not only what students will be learning but how you want them to engage with the work. Considering how to engage with new ideas should be informed by the students' current learning trajectory. Some students may show minimal facility with science concepts and vocabulary. Other students may be advanced in their knowledge and application. A preassessment provides teachers with valuable information about how prepared students are for new learning and should then inform future instructional choices.

If a preassessment provides information that indicates a classwide gap in knowledge, the teacher might include activities related to chunking that information, even if all the students have taken the prerequisite course and should already have attained proficiency. On the other hand, if a preassessment indicates a wide range of knowledge and skills, the teacher might differentiate student groupings and tasks to best account for these disparities. Student groupings may pair students of similar skill and ability to allow the teacher to provide scaffolded support and additional opportunities. The teacher also might decide to employ mixed groupings so students of varying proficiencies can work together to achieve a collective proficiency.

Effective science preassessments come in two types: (1) checking for understanding of background knowledge and skills that are important for the upcoming learning, and (2) checking for what students already know about the new concept. Sometimes the background skills and knowledge are about science topics, but they could be about general academic skills as well. Consider a chemistry lab comparing the concentration of an acid and the reaction rate. Students would be required to create a graph representing their results and then determine the line of best fit. It's important to determine before the activity if all students are able to create an accurate graph and also use their data points to determine the best fit line. A preassessment could consist of a group activity with sample data that requires each student to create a graph and then determine the slope of the best fit line. Students who are unsure of how to complete this activity would need extra time and support so creating the graph does not impede their ability to process the chemistry concepts.

In terms of science background knowledge, another good example is learning about the *Krebs cycle*, which is a "series of reactions responsible for most of the energy needs in complex organisms; the molecules that are produced in these reactions can be used as building blocks for a large number of important processes" (ScienceDirect, n.d.). This complex process involves multiple chemical reactions and can be very daunting to learn. At the heart of the cycle is the core concept of energy and energy transfer. Before beginning a lesson on the Krebs cycle, it's important to assess students' knowledge about the definition of energy, the law of conservation of energy, and how energy can transfer through a system. Without these basic understandings of how energy works, the Krebs cycle will simply be an exercise in memorizing steps.

In this case, teachers might use a group huddle to provide students the chance to engage with the needed background information. A group huddle ideally involves table groups of three to four students who record their thoughts on butcher paper or a large whiteboard (huddleboard). The teacher presents the complete Krebs cycle diagram on the board or with a video projector. In groups, on the butcher paper or whiteboard, students each record everything they know about the topic as well as questions about anything they don't understand or would like to know in upcoming lessons. This activity helps students build context for future learning and provides the teacher with important information about what students need to relearn before the Krebs cycle instruction can begin.

Presenting Content in Small, Sequentially-Related Sets

Chunking makes learning complex information manageable. After a student learns smaller chunks of information, these chunks can be gradually connected to others to encourage a holistic understanding of a scientific concept. Teaching in these bite-sized modules also leads to greater student facility with comparing and contrasting skills and sequencing discrete events later in the learning process (Koch, Philipp, & Gade, 2006).

In the process of introducing increasingly complex sequentially related concept sets, the teacher must continuously monitor student understanding. Although some cognitive dissonance is expected, the teacher should prepare to address student misconceptions before adding new concept sets or expanding the current ones. Teachers must remember that students are working to retrieve existing information while incorporating new ideas. This work is cognitively difficult, requiring effortful retrieval and interleaving topics, which research shows leads to enduring understandings and learning that sticks (Brown, Roediger, & McDaniel, 2014).

For example, consider a multistep process commonly taught in both high school biology classes and middle school life-science classes—the water cycle. The water cycle describes how water changes states of matter and circulates in and through our biosphere. Prior to understanding the water cycle, students need to access procedural knowledge related to the behavior of the water molecule itself. Students might begin through a study of phenomena with which they are familiar. In addition, it may be important to scaffold the activity with clues about sequence and relationships. Figure 3.2 provides a framework for engaging students in a discussion of water prior to attempting to build a complex understanding of the water cycle itself.

Water: Changes in States of Matter				
State of Matter	**Weather Example**	**Temperature**	**Speed of Molecules**	**How It Can Change to Another State**
Solid		Less than 32 degrees F (or less than 0 degrees C)		
	Rain		Medium speed	
Gas				When colder, changes into a liquid

Figure 3.2: Accessing prior knowledge about water.

As students show proficiency with smaller sets, it may be appropriate to expand them with new sets. In the process of expanding sequential sets, it's important to encourage transfer of acquired knowledge from prior sets. Teachers can use a chart like the one in figure 3.2 to preload prior knowledge from the last activity before introducing the four new vocabulary words related to the water cycle: *evaporation*, *condensation*, *precipitation*, and *collection*. In this example, when students add their prior knowledge to the diagram in figure 3.3, they construct explanations that lead to a fundamental understanding of new vocabulary words (which they will later write on the arrows) before the teacher even introduces related vocabulary words. In this way, students apply previously learned concepts that actually help define the four major processes illustrated in figure 3.3. In other words, they build the explanations prior to the introduction of the terms. Chunking provides a way to build understanding from the bottom up rather than asking students to memorize and consume information in a top-down fashion.

Figure 3.3: Water cycle diagram.

Further expansion of sets can lead to even more complex categorical comparisons. Figure 3.4 shows puzzle pieces representing important features of the four biogeochemical cycles (water cycle, nitrogen cycle, phosphorus cycle, and carbon cycle). These pieces can be cut up and randomly mixed in envelopes for student table groups to manipulate. On a poster board, digital whiteboard, or table surface, the teacher might ask students to categorize the pieces in labeled columns (see figure 3.5). In this activity, some pieces may fit in more than one category, so the teacher would need to be active in directing students through various possible correct ways of organizing the pieces. In addition, the activity lends itself to sequencing the pieces within the category columns via flowcharts and word webs.

Nitrogen Fixation	Condensation	Proteins	Atmosphere	Amino Acids	Rocks
Animals	Trees	Soil	Temperature	CO_2	Urine or feces
Sunlight	Decomposer	Plants	Precipitation	Run-off	Grass
Wind	Rocks	Dead plants and animals	Decomposer	Fossil fuels	Oceans
Nitrogen-fixing bacteria	Evaporation	Cellular respiration	Photosynthesis	Synthesis of amino acids	Nitrogen (N_2)
Ammonia (NH_3)	Transpiration	Evaporation	Combustion	Plants	H_2O
Rivers	Cell membranes, teeth, and bones	Percolation	Acid rain	Global warming	Lakes
Very slow process	Hurricanes and earthquakes	Does not cycle through the atmosphere	Sediment and rocks	Erosion of minerals	Legumes (beans, peanuts)
CFCs chlorofluorocarbons	Phosphorus cycle	Nitrates and nitrites	Groundwater	Plants	Microbes
Run-off	Fossil fuels	Oceans	Atmosphere		

Figure 3.4: Biogeochemical cycle puzzle pieces.

Water Cycle	Nitrogen Cycle	Phosphorus Cycle	Carbon Cycle

Figure 3.5: Biogeochemical cycle categorization grid.

Throughout the process of introducing more expansive sets, each set must logically build on the prior set in an iterative manner. Chunks must be both internally and externally consistent in order to build comprehensive student understanding. In addition, the rigor and complexity of the sets must be grade appropriate and align with the goals of the course.

Allowing for Processing Time Between Chunks

Before connecting or expanding each small, sequentially related set to other sets, students should engage in dedicated think time (processing time) both individually and in groups. Essential questions are a great way to allow for this type of processing. *Essential questions* are open-ended questions that typically focus on the *how* and *why* of a topic and allow for a variety of explanations (McTighe & Wiggins, 2013). Good examples of essential questions in science include: "What makes objects move the way they do?" "How are structure and function related in enzyme-mediated reactions?" and "Is aging a disease?" A good essential question related to the examples given in the prior section might be "What makes the weather change?" When placed between the set related to states of matter and the water cycle, students can access prior knowledge and make connections between the two sets.

When proposing an essential question to students, a good strategy is to give each student a notecard and ask him or her to respond in writing by the end of a short, defined period of time (typically five to seven minutes). Once the thinking has been captured on the notecards, the teacher has many options. The teacher may ask students to trade cards and then begin a class conversation. As the teacher calls on students, they'll be less reluctant to share because the card is someone else's thinking. As an alternative strategy, the cards may be categorized on a whiteboard (with magnets) according to different theories or concepts to help students visualize all the different types of responses. When intentionally designing lessons to incorporate student processing time, it is critical to capture their thinking in a visible way that can be shared with the class and contribute to an evolving group-level or classroom-level understanding (Ritchart, Church, & Morrison, 2011).

For example, the previously mentioned essential question "What makes the weather change?" might elicit a wide array of student responses with varying levels of complexity. Answers from high school students could be organized into categories like the Earth's revolution around the sun, temperature changes, atmospheric pressure changes, ocean currents, humidity, cloud coverage, deforestation, and others. At an elementary school level, the categories might be simplified as sun, wind, and seasons. Students' answers will be somewhat unpredictable, but a savvy teacher finds ways to honor each sample of student thinking to collaboratively build a collective understanding.

Element 7: Processing Content

When students pause between chunks, as previously described, science teachers must create opportunities for students to process each part to help ensure they understand and effectively use what they have learned. If science teachers forego dedicated situations to engage students in meaningful activities that help them analyze and process the material, they shortchange student learning. Making time for activities that allow for the processing of scientific content is essential in supporting the development of individual student insights. Direct instruction isolates each step in any form of procedural scientific knowledge and helps students build a comprehensive understanding of any given phenomenon.

Science teachers should develop student ability to process content because it helps them articulate their thinking, build conceptual understandings, and build scientific perspectives that draw thoughtfully from the available evidence. In addition, processing content builds capacity to engage in scientific discovery and problem-based learning. The strategies that follow provide a variety of compelling approaches to processing content. Figure 3.6 presents the self-rating scale to guide teachers in implementing this element.

Score	Description
4: Innovating	I engage in all behaviors at the Applying level. In addition, I identify those students who are not using the processes to better understand the content and design alternate activities and strategies to meet their specific needs.
3: Applying	Along with adequate guidance and support, I engage students in activities that help them process new information by making predictions, summarizing, and asking clarifying questions, and I monitor the extent to which students are using the processes to better understand the content.
2: Developing	I engage students in activities that help them process new information by making predictions, summarizing, and asking clarifying questions, and I provide adequate guidance and support.
1: Beginning	I engage students in activities that help them process new information by making predictions, summarizing, and asking clarifying questions but do not provide adequate guidance and support, such as modeling the use of these processes and providing students with adequate time to engage in these processes.
0: Not Using	I do not engage students in activities that help them process new information by making predictions, summarizing, and asking clarifying questions.

Figure 3.6: Self-rating scale for element 7—Processing content.

It is imperative to employ effective strategies so students can aptly process and internalize the material to facilitate their understanding. In this section, we focus on these strategies relevant to this element.

- Perspective analysis
- Thinking hats
- Concept attainment

Perspective analysis and thinking hats (de Bono, 1999) are two strategies that encourage students to thoughtfully explore multiple viewpoints related to scientific concepts, phenomena, and a variety of other scientific content. They are most effective when students read a variety of scientific texts and synthesize the information. Through this process, students realize that science is not simply an accumulation of facts but that scientific evidence forms the basis of hypotheses, theories, and general argumentation. Students use perspective analysis and thinking hats to identify, examine, and evaluate conflicting viewpoints. In addition, these strategies are extremely helpful in relating scientific material to students' lives and real-world scenarios.

Perspective Analysis

Perspective analysis encourages students to process scientific information by identifying different viewpoints applied to real-world examples. This is a skill that many teachers may be familiar with on standardized exams such as the ACT or SAT (for example, science "opposing viewpoints" passages in the reading section). Perspective analysis teaches students that science is constantly updated and moves forward through a constant clash of ideas. By coaching students through five process steps outlined in figure 3.7 (page 40), teachers provide a scaffold for students to explore a scientific topic or idea through multiple lenses (de Bono, 1999). In this exploration, students make value judgments and reflect on those judgments. Figure 3.7 shows elementary, middle, and secondary school examples. In the figure, note that the grade-cluster examples align with the generalized process steps and example questions in that column.

Perspective Analysis Examples			
Process Steps	Elementary School (Falling Objects)	Middle School (Global Warming or Climate Change)	Secondary School (Evolutionary Theory)
1. Identify your position on a debatable topic. *What do I believe is true about this topic?*	What do I believe about how fast objects fall?	What do I believe about the contribution of human activities to global warming?	What do I believe about the ability of species to change over geological time?
2. Identify your position on a debatable topic. *Why do I believe that this is true?*	Why do I believe that heavier objects fall faster?	Why do I believe that humans are the major contributor to increasing global temperatures over the past two hundred years?	Why do I believe that species have changed over time due to natural selection and evolution?
3. Identify the opposing position or positions. *What is another way of interpreting the evidence?*	What if objects fall at the same speed?	What is the position of those who believe that the earth's warming is not human initiated?	What are the alternative theories that explain the earth's great biodiversity?
4. Describe the reasoning behind the opposing position. *Why might someone else hold a different opinion or draw a different conclusion?*	What is the explanation or set of conditions that would show objects falling at the same speed?	Why do climate change deniers believe that global warming is a natural phenomenon?	Why have some people believed in different theories like Lamarck's acquired characteristics, Gould's punctuated equilibrium, or Darwin's theory of evolution?
5. When you are finished, summarize what you have learned. *What have I learned about this concept or phenomenon?*	What have I learned about falling objects and their relationship with speed?	What have I learned about the opposing viewpoints related to global warming and climate change?	What have I learned about the theories describing change over time?

Source: Adapted from Marzano, 2016.

Figure 3.7: Perspective analysis examples.

Thinking Hats

Science teachers can also coach students to use thinking hats (de Bono, 1999) to examine multiple facets of scientific concepts or phenomena. Thinking hats (each represented by one of six colors) provide students with mental lenses to apply to particular problems or positions.

- White (facts or evidence)
- Black (cautious)
- Yellow (benefits)
- Red (emotions)

- Green (creative ideas)
- Blue (conclusions)

Similar to perspective analysis, the thinking hats strategy helps teachers design questions and tasks for students. This strategy also helps organize approaches to a scientific issue via color-coded descriptions that act as different lenses to view a single idea or phenomenon.

Figure 3.8 illustrates a helpful lead-in to a class discussion surrounding cellular respiration and energetics using the thinking hats strategy.

Thinking Hats		
Description	**Examples**	
	Addressing a Problem	**Determining a Position**
1. White (Facts or Evidence)	What is the evidence?	What is the evidence regarding the effectiveness of exercise to reduce body weight?
2. Black (Cautious)	What is unknown, not well known, or potentially dangerous?	Can exercise plans lead to unintended consequences? Which studies highlight these?
3. Yellow (Benefits)	What is clear or well supported?	Which studies show a significant impact of exercise plans on body weight?
4. Red (Emotions)	How does this make me feel?	How do I feel about plans that only address body weight and not overall health?
5. Green (Creative Ideas)	What could be changed or what are the alternatives?	What are the factors that relate to body weight and how are they correlated with overall health?
6. Blue (Conclusions)	Why did it happen this way or what is the potential application?	Do exercise plans work? What is your recommendation to people looking to maximize personal health?

Source: Adapted from de Bono, 1999.

Figure 3.8: Thinking hats example.

To facilitate an effective conversation, the teacher might choose to give each table group a selection of studies or abstracts for some prereading. In addition, the teacher expands the activity to incorporate some independent research related to the first three hat colors.

A variation on this theme may also be applied to laboratory investigations as a way to organize student lab journals. Although thinking hats are metaphorical tools, elementary teachers may wish to make this more tangible by creating different-colored paper hats that students may wear when engaged in specific thinking processes. For example, the teacher can say, "Let's put on our white hats to examine the data at your tables."

Teachers can utilize perspective analysis and thinking hats in a variety of contexts to support individual work and students working in pairs or small groups. In addition, the different lenses in either strategy may be delegated throughout the classroom and later coalesced via jigsaw structures, of which you will see more in element 22 (page 93). If every individual is completing every portion of one of the strategies, the teacher may later group these individuals to compare and contrast their answers in order to broaden understanding and gain new perspectives. Students can also apply either strategy to evaluate a single source of scientific information (for example, journal article, newspaper article, YouTube video) to determine if there is sufficient evidence given to support the position. Teachers can modify both strategies to support a variety of goals related to scientific literacy, such as source validity, reading comprehension, and vocabulary development.

Concept Attainment

In concept attainment, teachers create structures for students to struggle through some cognitive dissonance in order to construct deeper understandings. *Concept attainment* is the result of students comparing and contrasting both examples and nonexamples for a particular scientific idea. Concept attainment involves students in the learning process and asks them to make meaning while observing examples and nonexamples (Dean, Hubbell, Pitler, & Stone, 2012). In this sense, the strategy is generative, as students build a comprehensive understanding from select pieces. Although similar to the chunking process mentioned earlier, concept attainment should intentionally incorporate misconceptions so students can learn to discriminate these from legitimate concepts. This section presents methodologies that science teachers may use to lead pair or small-group concept attainment exercises to help students clearly understand a complex topic and eliminate misconceptions.

Cartoon analysis can be an effective way to help students discriminate between legitimate scientific ideas and those that are false or misleading. Scientific misconceptions are persistent if not directly addressed in the classroom. For example, when teaching evolutionary theory at the middle or high school level, teachers can present cartoons with embedded questions to students and poll the class using notecards (Anderson & Fisher, n.d.; Anderson, Fisher, & Norman, 2002). On the notecards, students answer the questions and include the answer (A, B, C, or D), along with an explanation. The teacher then collects the cards, shuffles them, and has small table groups of three to four students review the answers and explanations to see if they can arrive at the correct answer and debunk the others. Questions for a cartoon about natural selection might include the following:

> When there is not enough food, what happens?
>
> a. All the ducks work together to find food. They cooperate with each other!
>
> b. No, some ducks need to change so they can learn to eat other foods.
>
> c. Some ducks won't get enough food to eat and may get sick and weak.
>
> d. Ducks always fight for food, and the biggest, strongest ones always get the most.
> (Anderson & Fisher, n.d.)

In the context of this particular cartoon, students should uncover many misconceptions about evolution: evolutionary fitness is not defined as large organism size or great strength (it is defined as the ability to survive long enough to reproduce); individual organisms cannot adapt within their lifetimes (only populations can adapt over many generations); and cooperation is rare in the animal kingdom (while competition for resources predominates; Anderson, Fisher, & Norman, 2002).

Comparison charts and *multiple-matching activities* require students to identify, compare, and contrast examples and nonexamples of a particular concept. Figure 3.9 shows a comparison chart for the concept of gene regulation in bacteria via operons. An *operon* is a "group of closely linked genes that produces a single messenger RNA molecule in transcription and that consists of structural genes and regulating elements (such as an operator and promoter)" (Merriam-Webster, n.d.).

Characteristic or Feature	Repressible Operon	Inducible Operon
Definition		
Normal operon condition (ON or OFF)		
Relationship to metabolism		
Repressor protein produced is . . .		
Example		
Type of gene regulation (positive or negative)		
Draw each when the operon is turned on. Include the repressor protein.		

Figure 3.9: Operons comparison chart example.

Figure 3.10 shows a multiple-matching activity requiring students to discriminate operon-related vocabulary and apply the terms to specific scenarios. With multiple matching, students apply one letter or many, depending on the context. Multiple-matching activities are best followed by a teacher-directed class discussion to focus students on the many possibilities for each scenario. In regard to concept attainment, this assures that students are able to fully incorporate all the elements of the concept that may have been chunked out in an earlier strategy.

Match the phrase or sentence on the left with the correct term or terms on the right. Choices may be used once, more than once, or not at all. *Each question may require more than one letter to be correct.*

_____ 1. Binds to promoter

_____ 2. Binds to operator

_____ 3. Regulates anabolic pathways

_____ 4. Regulates catabolic pathways

_____ 5. Produces tryptophan

_____ 6. Makes enzymes necessary for metabolism

_____ 7. Makes repressor protein

A. Feedback

B. Operator

C. Promoter

D. Gene cluster

E. *lac* operon

F. *trp* operon

G. Regulator gene

H. Inducer

I. Co-repressor

J. RNA polymerase

K. Repressor

Figure 3.10: Multiple-matching activity example.

Element 8: Recording and Representing Content

The final element of direct science instruction asks that students have an opportunity to record and represent the new content in ways that are personally meaningful. The goal is for students to "create an internal representation of the content" (Marzano, 2017, p. 32). These internal representations may be conducted in a manner that is linguistic or nonlinguistic. When recording or representing content, it is critical to engage each individual student's creative impulses, ensure that individual student voices are authentically represented, and develop activities that are both unexpected and memorable. The strategies provided help students because they are novel, memorable, and provide hacks that make learning easier. Some of the strategies in this element (specifically, dramatic enactments and mnemonic devices) can help you create unforgettable learning experiences for students. These strategies in particular maximize the power of moments. A defining moment in the classroom is characterized by one or more of the following: elevation, insight, pride, or connection (Heath & Heath, 2017). When teachers ask students to interact with the content by bringing in aspects of their personality (or when teachers do the same), the learning becomes a socially constructed phenomenon tied to personal identity and pride.

Figure 3.11 presents the self-rating scale for this element.

Score	Description
4: Innovating	I engage in all behaviors at the Applying level. In addition, I identify those students for whom the process does not help them discover and remember new distinctions they have made about the content and design alternate activities and strategies to meet their specific needs.
3: Applying	Along with adequate guidance and support, I engage students in activities that help them record and represent their thinking regarding new content and monitor students to ensure that their actions help them discover and remember new distinctions they have made about the content.
2: Developing	I engage students in activities that help them record and represent their thinking regarding new content and provide adequate guidance and support.
1: Beginning	I engage students in activities that help them record and represent their thinking regarding new content but do not provide adequate guidance and support, such as allowing and encouraging students to record and represent their thinking in ways that are most comfortable for them individually, modeling the different ways to record and represent their thinking, and providing adequate time to record and represent their thinking.
0: Not Using	I do not engage students in activities that help them record and represent their thinking regarding new content.

Figure 3.11: Self-rating scale for element 8—Recording and representing content.

This section provides specific examples geared toward teaching these strategies within element 8.

- Summaries
- Free-flowing webs
- Dramatic enactments
- Mnemonic devices

Summaries

Summarizing is a key strategy in the science classroom to condense detailed webs of information into a format that communicates the main ideas succinctly. As students sift through data and concepts, they must exercise their critical thinking skills. In many data sets, there are outliers or data that are considered insignificant. In addition, many scientific concepts involve various levels that may not be important to learn, depending on the grade level or stage of learning in the year. In some studies, summarization strategies show little utility because most students simply create nondiscriminating lists (Dunlosky, Rawson, Marsh, Nathan, & Willingham, 2013).

A major benefit of summarization strategies is that they help students focus on the core idea or a main conclusion. In this way, students should be able to separate the essential from nonessential information to construct meaning. A great variety of summarization strategies are at the teacher's disposal and can be employed at almost any grade level. In the science classroom, the goal of a summary should be to make the information accessible to the layperson. In other words, a summary should not simply regurgitate vocabulary but explain in a way that is enlightening to someone who has not studied the topic.

An *elevator pitch* can be an effective way for students to construct an oral or written summary. In an elevator pitch, the student must create an explanation that lasts no longer than thirty seconds to two minutes (the length of an elevator ride). The pitch should address both the *how* and the *why*. In a high school physics classroom, for example, this activity might involve the interrelated concepts of mass, weight, and gravity. After some initial preparatory reading, the teacher asks students to create an elevator pitch to explain how these ideas could be applied to the same object on two different celestial bodies, for example the Earth and its moon. In order to avoid unnecessary stress for students and to allow think time, encourage students to write these out (preferably on notecards). After writing, teachers can give students a short amount of practice time to rehearse and revise their pitches and to time themselves. Students can then share their pitches orally in small groups with the teacher using some as exemplars in whole-class discussion.

Using social media conventions familiar to students are effective in helping students summarize. *Tweets* (short social media posts) and *hashtags* allow students to be extremely creative. In both conventions, students are forced to take complex ideas and distill them down into a short explanation (a tweet can contain up to 280 characters) or summary phrase (a hashtag typically is no longer than a standard sentence). Most middle school and high school students are digital natives fluent in these patterns of summarization in their social lives. They can apply these skills productively to encourage synthesis and deep learning.

For example, in the chemistry classroom, students must understand several difficult and interrelated concepts in order to gain a comprehensive understanding of the basics of organic chemistry: electronegativity, polarity, bond formation, valence shells, intermolecular forces, and more. To help students learn these kinds of concepts, a teacher might share an instruction sheet that tasks student groups with becoming experts on a particular topic. Each group researches the topic and composes a tweet on a common shared document or presentation slide. Figure 3.12 (page 46) represents a teacher-generated example of such a tweet slide. Once the slide deck is completely populated and the teacher reviews all slides for validity, it becomes a shared class resource that students can use as a study guide or to direct further peer-to-peer instruction (especially if students all have their own devices).

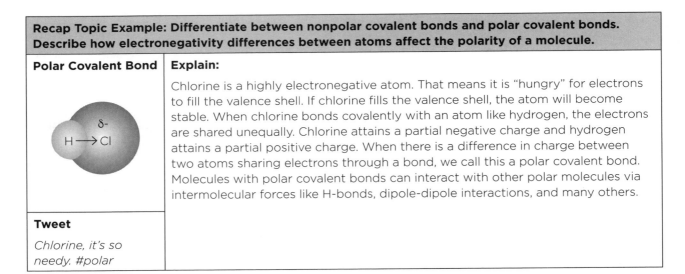

Recap Topic Example: Differentiate between nonpolar covalent bonds and polar covalent bonds. Describe how electronegativity differences between atoms affect the polarity of a molecule.

Polar Covalent Bond	Explain:
	Chlorine is a highly electronegative atom. That means it is "hungry" for electrons to fill the valence shell. If chlorine fills the valence shell, the atom will become stable. When chlorine bonds covalently with an atom like hydrogen, the electrons are shared unequally. Chlorine attains a partial negative charge and hydrogen attains a partial positive charge. When there is a difference in charge between two atoms sharing electrons through a bond, we call this a polar covalent bond. Molecules with polar covalent bonds can interact with other polar molecules via intermolecular forces like H-bonds, dipole-dipole interactions, and many others.
Tweet	
Chlorine, it's so needy. #polar	

Figure 3.12: Teacher example of tweet slide.

In addition to the aforementioned strategies, students may also summarize with *graphical models, top-ten lists,* and *single-word summaries.* Students may be asked to create graphical models that include no text (or very limited text) to express a relationship—such as enzyme-substrate interactions and the factors that influence them. These unlabeled models may also be used in a pair-share activity so students can attempt to interpret the models of other students. Top-ten lists ask students to create a list of the top ten things that are most important to remember for a particular concept or idea. These lists are particularly effective when asking students to summarize learning that occurs over a long span of time, such as the end of a multi-week unit on cellular respiration. Finally, single-word summaries are effective in helping students build facility with vocabulary, as they provide simplified definitions that can serve as scaffolding to build toward more complex definitions. For example, students might struggle with a term like *anabolism.* If students can self-construct the term *build* as a single-word summary, they now have a foundational understanding and can later add more layers to that understanding: net input of energy, formation of a higher potential energy product, and more.

Free-Flowing Webs

Free-flowing webs, also known as *concept webs* or *word webs,* require students to place big ideas in central shapes, such as ovals, and then connect them to other shapes with connecting lines. Ideally, the connecting lines have directionality (arrows) and a short descriptor placed over the connecting line describing the relationship. The directionality of the arrow indicates cause and effect or that the prior term supports the latter term. The descriptor over the arrow provides a very brief description of the relationship between those two terms.

Figure 3.13 shows an example of a free-flowing web related to DNA replication.

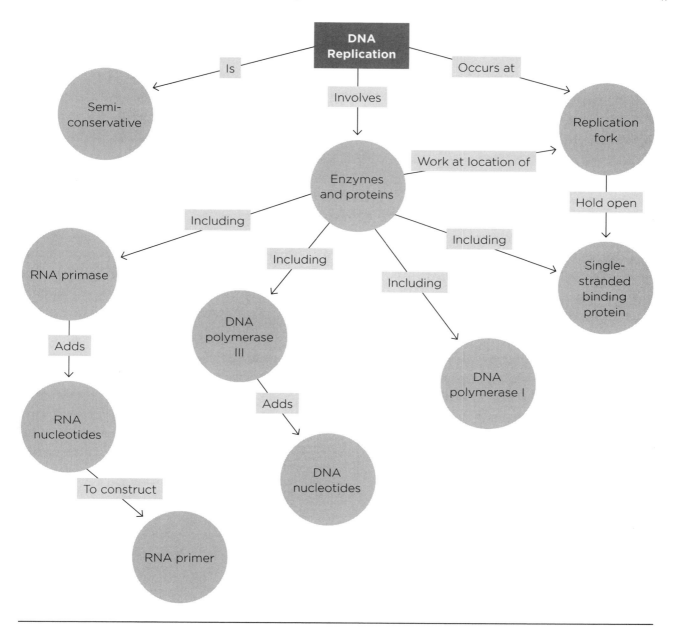

Figure 3.13: Free-flowing web example of DNA replication.

When creating a free-flowing web, it is important that students begin with a small grouping of topics so the relationships don't become cluttered. They should include only the most essential pieces of information to convey fundamental relationships. As students gain facility, the teacher may coach them on how to connect multiple groupings. In class, free-flowing webs are best completed in groups on huddleboards (collaborative whiteboards) so students may easily erase and revise. Although the web product may be useful for students as a study guide, the process of constructing the web is the most valuable part of the activity. In fact, with many webbing activities, it may be beneficial to have students begin a web as a group at a table and then rotate the groups to other tables to revise the webs and add new elements to what other groups have begun. Through this process, students collaborate with the entire class during the rotation, and the webs become effectively self-correcting.

Dramatic Enactments

Dramatic enactments involve students role-playing elements of specific scientific processes. When developing dramatic enactments, students must create a script, assign roles, enact the roles, and ensure that the dramatic enactment is coherent and a faithful rendering of the source material. Throughout this process of producing the dramatic enactment, students make choices about what is essential, what can be left out, and how to best communicate the main ideas.

These enactments can be very simple and involve one or two students. They also may become very complex and involve an entire class, with each student in a unique role. In role play, engagement levels are high because students must be aware of their parts while also being cognizant of how they relate to other roles. Research shows that dramatic integrations show strong effects on student achievement when they are led by a classroom teacher (and not an outside artist), span multiple lessons, and are integrated into the science curriculum (as opposed to other subject areas; Lee, Patall, Cawthon, & Steingut, 2015). However, dramatic enactments can cause students some stress if they are not outgoing or extroverted. As a teacher, it is critical to show understanding of these students and do everything possible to make the environment a safe space for them while communicating that mistakes are part of the fun.

A simple dramatic enactment might involve a small group of students in the middle of the classroom. A biology example could include an enactment of the role of adenosine triphosphate (ATP) in glycolysis. Six student volunteers may be assigned the role of glucose (joining hands to form a hexagon). Two other volunteers could play the roles of ATP molecules. The teacher could give three stickers to each ATP volunteer, each labeled with *P* for *phosphate*. While describing how dephosphorylation and phosphorylation transfer stored energy, the teacher coaches ATP molecules to lose a phosphate and transfer them to the student hexagon (glucose). Phosphate transfer then causes the glucose molecule to become highly reactive and unstable, breaking into two chains of three students. In the course of this kind of enactment, the rest of the class (audience) can question or direct the performing students' actions.

A more complex enactment might engage a larger group of students and require them to write the script, assign roles, make props, and perform for an audience. Figure 3.14 shows an AP biology assignment called Immunothespians, in which half the students were assigned the humoral (antibody-mediated) immune response and half were assigned the cell-mediated immune response.

Immunothespians			
Tasks: 1. In a group (ten to fourteen students), you will be in charge of acting out your assigned immune system pathway. 2. Highlight your assigned target:			
VII	Describe the events of a **humoral** (antibody-mediated) response pathway. Explain the process using the following terms: **B-cell receptors** called **antibodies** (a.k.a. **immunoglobulins-Ig**), **B-cells**, **plasma cells**, **macrophages**, **lymphokines (cytokines)**, **MHC**, **antigens**, **self**, **non-self**, and **memory B-cells**.	35.2, 35.3	Class notes, lecture, target packet
VIII	Describe the events of a **cell-mediated** response pathway. Explain the process using the following terms: **Helper T-cells, killer (cytotoxic T) cells, T-cells, macrophages, MHC, antigens, self, non-self, lymphokines (cytokines),** and **memory T-cells.**	35.2, 35.3	Class notes, lecture, target packet

3. On the backside of this page, you have a list of thirty-three phrases (some are distractors and DO NOT APPLY TO ANY PATHWAY). Determine which phrases go along with YOUR target and get them arranged in the correct order.

4. Once you have figured out your general script, be sure to add any necessary information outside the phrases (like *apoptosis*, and so on). Also, think of specific examples or scenarios to include in your performance. **Make connections.**

5. Decide on roles for the performance. Who will play what parts? How will you distribute the lines? **Everyone must have a role and a speaking part.**

6. Which props will you use? How will you make it entertaining and informative?

7. Your performance should be about **ten minutes** in length and will be followed by a five-minute Q&A with the audience.

Grading (16 points total):

Content accuracy of skit	5	4	3	2	1
Unique examples, connections, and creativity			3	2	1
Effectiveness of props			3	2	1
Ability to answer questions and provide feedback			3	2	1
Involvement of all group members				2	1

The people evaluating you will have a key of the phrases in the correct order, so they will know how it is supposed to work!

Figure 3.14: AP biology assignment example.

continued →

Immunothespians

Bank of Key Phrases:

Helper T-cell (T_H) proliferates, secretes IL-2

Chemokines released

Macrophages arrive

Tissue heals

Cytotoxic T-cells (T_C) releases perforin

Virus infects cell

Class 1 MHC molecule binds to viral antigen

Viral antigen transported by class 1 MHC molecule to cell surface

Antigen presented to T_C

CD8 enhances binding

T_H cell also stimulates T_C cell with IL-2

Infected cell lyses

Injured cells release histamine

Capillaries dilate

Class 2 MHC molecule binds to foreign peptide

Macrophage ingests a pathogen, or B-cell membrane receptor binds to an antigen

T_C is activated

Antigen transported by class 2 MHC molecule to cell surface

Antigen is presented to helper T-cells

Memory B-cells have membrane antibodies, remain in blood

IL-2 and activated T_H cell activate B-cell

B-cell proliferates

Antibodies dispose of antigen

Antibodies bind to antigen

Complement proteins present in serum link antibodies

Complement proteins are activated, inflammation occurs

Other proteins activated, cascade reaction

Membrane Attack Complex is activated

CD4 enhances binding

Pores form in membrane

Cell lyses

Plasma B cells secrete antibodies

Neutrophils arrive

In this activity, students received a list of thirty-three phrases that applied to one pathway or the other (or both). However, students were required to use their resources to determine which phrases applied and which didn't apply, and how to sequence the events. Since the performance itself can sometimes be messy and include humorous moments, the teacher must be cognizant to include a question-and-answer session after the performance. In addition, the strategy works especially well if the audience is tasked with evaluation. The teacher can coach the audience to look for both exemplars and misconceptions. If implementing these more complex enactments, time included for post-performance class discussions should be at least as long as the performances themselves.

Mnemonic Devices

Mnemonic devices are invaluable tools in helping students remember, record, and represent critical content (Marzano, 2017). In the science classroom, mnemonic devices are especially helpful for remembering how multiple factors or concepts relate to other concepts. Mnemonic devices based on common, everyday words can be quite helpful, but strange and silly ones are often the most memorable for students. A simple example from a biology classroom expresses the relationship of the highly electronegative atoms to polarity: *SON = Polar.* Sulfur, oxygen, and nitrogen are common highly electronegative atoms that make organic molecules polar (or parts of organic molecules).

In the chemistry classroom, the teacher could instruct students to remember the diatomic molecules by remembering the acronym *HOFBrINCl* (which stands for the atoms *hydrogen, oxygen, fluorine, bromine, iodine, nitrogen,* and *chlorine*). This also helps give the mnemonic device a backstory. The teacher may develop a story about a fictional lonely professor named Hofbrincl and describe how he is terrified to be alone. Segueing from this story, the teacher may discuss the atoms *hydrogen, oxygen, fluorine, bromine, iodine, nitrogen,* and *chlorine*. Like Professor Hofbrincl, the teacher explains how these atoms are particularly lonely and can never exist as individual atoms (H, O, F, Br, I, N, Cl)—they must always be together when not part of a compound to become diatomic molecules (H_2, O_2, F_2, Br_2, I_2, N_2, and Cl_2). Once students understand the story and the acronym, they have a memorable mechanism to identify these chemical elements (H, O, F, Br, I, N, Cl) and make the appropriate diatomic molecules (H_2, O_2, F_2, Br_2, I_2, N_2, Cl_2) in a balanced chemical reaction equation.

Mnemonic devices also might become more elaborate and nonsensical. A particularly effective mnemonic device works for teaching the structure of the DNA molecule. Students have a particularly hard time remembering the nitrogenous bases and their classifications. The following mnemonic devices are memorable and can help students with these relationships.

Cing Tut lies in a Single Pyramid

Pur Ag will Double your Wealth

Although superficially confusing and silly, these mnemonics make use of prior knowledge of Egyptian history, chemistry, and some general associations. Students quickly remember that C (cytosine) and T (thymine) are single-ring structures classified as pyrimidines. They also can recall that A (adenine) and G (guanine) are double-ring structures classified as purines. The unexpected and humorous can be effective tools in developing mnemonics.

GUIDING QUESTIONS FOR CURRICULUM DESIGN

This design question pertains to conducting direct instruction lessons: *When content is new, how will I design and deliver direct instruction lessons that help students understand which parts are important and how the parts fit together?*

Consider the following questions aligned to the elements in this chapter to guide your planning.

- **Element 6:** How will I chunk the new content into short, digestible bites?

  _____

  _____

  _____

  _____

  _____

- **Element 7:** How will I help students process the individual chunks and the content as a whole?

  _____

  _____

  _____

  _____

  _____

- **Element 8:** How will I help students record and represent their knowledge?

  _____

  _____

  _____

  _____

  _____

Summary

Within the second major category of *The New Art and Science of Teaching* framework—content—science teachers effectively use strategies within four areas in an intentionally coordinated way to help students learn the information and skills at the center of instruction. As noted in the introduction, these areas include direct instruction, practicing and deepening, and knowledge application lessons, plus strategies applicable to all these types of lessons. Direct instruction, the focus of this chapter, is essential when teachers introduce new learning to students in the science classroom, whether it be a new skill, process, strategy, or content information. Teachers then design and conduct lessons that allow students to practice and deepen their understanding of new learning.

CHAPTER 4

Conducting Practicing and Deepening Lessons

Practicing and deepening strategies are different for procedural and declarative knowledge. This is a very important distinction to keep in mind when considering the use of specific instructional strategies, as practice develops procedural knowledge while deepening strategies support declarative knowledge.

Procedural knowledge involves large, comprehensive processes and the basic skills and tactics that are the components of the larger processes. Certainly, engaging in the science and engineering practices from the NGSS qualifies as an example. Each science practice includes many embedded skills and strategies, such as making an accurate claim, generating (or citing) relevant evidence, and providing accurate and clear science reasoning that ties the evidence back to the claim. In addition, laboratory investigations allow students to engage in the NGSS practices in a hands-on manner. For students to be successful in these investigations, they must be coached in both the purpose of the investigations and the process of evidence evaluation (Millar, Lubben, Got, & Duggan, 1994; Osborne, 2014). Regardless of context, students must practice procedural knowledge to the point where they can execute it fluently and without significant error.

Declarative knowledge is informational in nature. As opposed to "knowing how," declarative knowledge has been characterized as "knowing that" (ten Berge & van Hezewijk, 1999, p. 605). It involves knowing details such as science facts and terminology but also more broad information about science concepts. Science learning across the country has, unfortunately, often focused too heavily on declarative knowledge as students learn facts, formulas, and theories. However, even procedural learning must weave in a fair amount of declarative knowledge. For example, to effectively engage in a scientific argument, students must understand the relevant science concepts, how they apply to a given situation, and the ways in which particular evidence supports a particular argument. Students must practice procedural knowledge, and they must deepen their declarative knowledge.

These elements pertain to practicing and deepening lessons.

- **Element 9:** Using structured practice sessions
- **Element 10:** Examining similarities and differences
- **Element 11:** Examining errors in reasoning

Element 9: Using Structured Practice Sessions

As noted, students need to practice procedural knowledge. Therefore, teachers should plan and conduct learning situations for students to practice the new skill, process, or procedure so they can apply it autonomously in a novel way. Continuous structured practice develops scientific and laboratory skills that are durable and transferable. With enough practice and guidance, students should gain the ability to engage in science with minimal support.

In the following sections, we recommend specific structured practice strategies (modeling, guided practice, and varied practice) to help students master scientific skills and content knowledge. Figure 4.1 presents the self-rating scale to guide teachers in addressing this element.

Score	Description
4: Innovating	I engage in all behaviors at the Applying level. In addition, I identify those students who are not systematically developing fluency and accuracy and design alternate activities and strategies to meet their specific needs.
3: Applying	I engage in activities that provide students with structured practice sessions without significant errors or omissions and monitor the extent to which students are systematically developing fluency and accuracy.
2: Developing	I engage in activities that provide students with structured practice sessions without significant errors or omissions.
1: Beginning	I engage in activities that provide students with structured practice sessions but do so with errors or omissions, such as assigning independent practice for which students are not adequately prepared and not providing enough practice over time.
0: Not Using	I do not engage in activities that provide students with structured practice sessions.

Figure 4.1: Self-rating scale for element 9—Using structured practice sessions.

We focus specifically on the following strategies aligned to this element that represent ways teachers can assist students in honing procedural knowledge.

- Modeling
- Guided practice
- Varied practice

Modeling

Prior to a hands-on activity, many teachers find that modeling the skill first is effective in providing both visual and oral cues so students can envision doing the task themselves. Modeling is a process that science teachers frequently use to teach lab-based skills and to use lab equipment. In nearly every science course, students use lab glassware and other science materials. Skillful practice in the use of science tools is necessary to prevent these tools from getting in the way of learning the science. To ensure students are safe and effective in learning science, teachers need to model effective practices. Teachers cannot assume that students have the background experience and knowledge to be safe using a Bunsen burner or effectively use a microscope. These skills must be explicitly taught through modeling. Through the strategy of modeling, teachers can show how to do each step while stating aloud their thinking.

For example, when students learn to use a microscope, there are various goals in the learning process, including carrying the microscope properly, knowing the microscope's parts, understanding when to use each of the objectives, being skillful in how to bring the subject into focus, and so on. During this process, it is helpful if students watch and listen and then individually practice. For example, the teacher can project an image from a microscope and then explain his or her thinking to students while modeling how to properly use the coarse or fine adjustment, which objective to choose, and so on.

Teachers can also model specific pitfalls in a science skill, such as not focusing on air bubbles in samples like pond water. Figure 4.2 shows a model of how this might happen for an introductory lesson on microscopy. Teacher clarity about science skills and the process they will use to model them is one of the values in creating an outline such as this; clarity for the teacher paves the way for a clear and helpful modeling experience for students.

Think Aloud (Teacher Says)	Model the Skill (Teacher Does)
Preview the task: The goal is for students to safely bring a variety of objects into proper focus for more up-close inspection. Students will: • Know the various types of microscope. • Be able to safely and correctly bring items into focus at the appropriate magnification. • Identify structures or processes that are taking place in their specimen.	• Prepare a handout that outlines the steps for using a microscope. • During the microscope overview, project the image using the document camera or show a video (many available online) showing the steps of using the microscope correctly. Prepare this in advance so I can narrate. In addition, create a handout that outlines the parts of the microscope and the steps we will follow in proper microscope use.
First, underscore the importance of being very careful when moving the microscope. Share the need to always have two hands on the microscope, with one on the arm and one on the base. In addition, explain that after carefully setting the microscope on the lab bench, do not slide it across the counter, since it will bounce and vibrate, potentially doing damage. During each step of the explanation, ask students to explain *why* each of these precautions or reminders is in place. Finally, explain how to plug in and turn on the microscope, asking students what the power source might be needed for.	• Show students how to properly grasp the microscope by the arm, remove it from the cabinet, and then slide the second hand under the base. • Demonstrate carefully walking across the room carrying the microscope. • Show students what happens when the microscope is dragged across the surface. Then ask students to explain what damage might be done in the process. • Plug in the microscope, and show students where the power switch is located.
Second, explain how the slide clips or other apparatus are used to hold the slide in place. Small movements are magnified greatly when using the microscope, so these are needed to hold the specimen in place. It's important for the slide to lie flat on the stage and be centered over the aperture. Finally, explain that when moving a slide on the stage, the movements will be opposite of the view through the microscope.	• Demonstrate (showing on the document camera or by having students gather around) how to use the stage clips (or other mechanism) to hold the slide in place. • Model how to move the slide on the stage, particularly if there is a mechanical stage with knobs. Handle the microscope slide using the stage clips and stage adjustment knobs (if present).

Figure 4.2: Model for microscopy lesson.

continued →

Think Aloud (Teacher Says)	Model the Skill (Teacher Does)
Third, share that choosing the right objective is critical to see what is intended. Explain that, just like when moving the slide on the stage, small movements have dramatic impacts. It is for this reason you can do damage to the high-power objective, meaning that you should always begin with the low-power objective. Next, explain the process of focusing on low power and then moving to high power, always beginning with the coarse adjustment, and the importance of only using the coarse adjustment with the lowest-power objective.	Begin by showing how close the slide is to the high-power objective to underscore the need to begin with low power. Next, demonstrate the use of the coarse adjustment knob, with the low-power objective in place and the stage at its closest point. This is needed to get close to the ideal focus. Then demonstrate the use of the fine adjustment, moving to higher powers (if needed), and then refining the focus by using only the fine adjustment. This is a key step to show via the document camera or recorded video so students can see this in action.

In elementary or middle school, teachers can also model the proper use of a balance to determine the mass of an object, whether that be with an electronic balance or a triple-beam balance. A triple-beam balance brings forward concepts that are helpful for students to visualize, such as the need for the balance to be in balance before adding any mass to the weigh pan, thereby ensuring the science tool is working properly. The teacher can then place the object on the pan and have that equal out with the sliders being moved. While doing so, the teacher models and explains his or her thinking: if it's really unbalanced, then move a larger weight. If it's only slightly unbalanced and bouncing back and forth, then the teacher can use a smaller mass to get to the final correct mass. This process provides a good opportunity to demonstrate how to read another science tool, record answers to the proper number of decimal places, and even how to estimate between the lines when the situation arises.

Finally, it has become very popular in schools to use data collection devices, such as probes, that connect to phones, iPads, computers, and so on. Some of these tools are becoming much more reasonable in cost, allowing for efficient data collection that can be sent to other devices. These tools provide a wealth of information, but students must understand how to use the probe as well as the digital interface that provides the data.

As noted earlier, it's important that these tools don't become a barrier to learning. This can include creating graphs that show carbon dioxide or oxygen levels, the amount of force that is exerted on an object over time, or the temperature of a given reaction over time. In order to prevent these tools from getting in the way of learning the science, teachers need to effectively model how to use them.

Guided Practice

Engaged learning often includes students working in cooperative groups. These collaborative learning experiences can serve as powerful learning moments when structured, supported, and implemented well. To be successful, it's important that science teachers teach students skills for how to interact with peers as well as doing the science. One of the best ways to conduct this form of guided practice and teach the collaborative skills that are so important in science learning is to use strategies such as Kagan cooperative learning strategies. Spencer Kagan and Miguel Kagan (2015) outline many strategies for various purposes in the classroom. For example, in the strategy called Rally Coach, a pair of students works together using one pencil and one problem set. Partners take turns, with one student writing and the other student listening, checking, coaching, and praising. Partners alternate roles as they move through the problems. A strategy such as this is particularly helpful when doing mathematics-related tasks such as generating Punnett squares, calculating the caloric energy in a particular food, or determining the velocity of an object.

Another strategy to easily apply to specific skills, such as measuring an object or bringing an object into focus on a microscope, is simply taking turns between two or three students. When students watch one another perform a task, they become aware of procedural mistakes their partners might make. With the right coaching, these moments benefit the student performing the skills as well as the student or students providing the feedback. When this process also includes a rotation to pick up where the other student was working, it forces close observation for a seamless transition.

A final guided-practice strategy includes supporting student writing or conversational skills with general question prompts that apply to a wide variety of science scenarios. These are particularly helpful when eliciting writing samples or to promote conversations in groups. Teachers can use question prompts for students to begin and end conversations so they remain focused on the topic and as well as provide a framework for communicating.

As students practice writing or speaking about their learning and understanding, they deepen their learning and become aware of ideas they are not as confident about. Sentence starters to promote this guided practice include the following:

- What are the steps you used to determine . . . ?
- What if we changed . . . ?
- How is this problem different or similar to . . . ?
- Is there another way to find . . . ?
- What evidence do you have to support or refute . . . ?
- What is the key to solving this problem . . . ?
- How does . . . compare and contrast to . . . ?

Varied Practice

A fundamental part of many science courses involves problem solving. Certainly, the arithmetic becomes more complex in middle school and high school. These problem tasks come in a variety of formats, requiring different types of thinking. For example, when students study chemical reactions and matter conservation, they learn to balance chemical equations. This involves an understanding of reactants, products, coefficients, subscripts, counting atoms, and determining correct ratios. This is an opportunity for students to engage in varied practice in several ways. We recommend doing bits and pieces of this multitopic and iterative approach with any science content learning, since interleaving practice with other learning is one of the most powerful learning tools in a teacher's toolkit (Brown, et al., 2014; Lang, 2021). Following are three possible ways to vary practice, using the example of balancing equations.

Choose Varied Levels of Challenge

In this approach, students choose from practice problems at various levels of challenge. It is not necessary for students who are quickly proficient at the task to grind through extensive repetition of simple problems. Similarly, it is not productive for students to flounder with problems that are beyond their current state of understanding. By providing a practice set that allows students to choose the level of challenge, they can be more in control of their learning and choose problems that are most appropriate for them. For example, ask students to accurately complete any combination of problems that total up to twelve points of practice, with easy problems worth one point, medium problems two points, and challenge problems four points. Choice allows students to satisfy the requirements of balancing equations, but in a way that best meets them where they are in their learning. Students can complete problems in class or as a homework assignment. Following are some example problems.

Easy problems (one point each):

1. _____ $P_4(s)$ + _____ $Cl_2(g)$ → _____ $PCl_3(l)$
2. _____ $Sn(s)$ + _____ $Cl_2(g)$ → _____ $SnCl_4(l)$
3. _____ $P_4(s)$ + _____ $O_2(g)$ → _____ $P_4O_{10}(s)$

Medium problems (two points each):

1. _____ $Mg(s)$ + _____ $NH_3(g)$ → _____ $H_2(g)$ + _____ $Mg_3N_2(s)$
2. _____ $Fe(s)$ + _____ $H_2O(g)$ → _____ $Fe_3O_4(s)$ + _____ $H_2(g)$
3. _____ $P_2O_5(s)$ + _____ $H_2O(l)$ → _____ $H_3PO_4(aq)$

Challenge problems (four points each):

1. _____ $C_2H_8N_2(l)$ + _____ $N_2O_4(g)$ → _____ $N_2(g)$ + _____ $CO_2(g)$ + _____ $H_2O(g)$
2. _____ $Ca_3(PO_4)_2(s)$ + _____ $SiO_2(s)$ + _____ $C(s)$ → _____ $P_4(s)$ + _____ $CaSiO_3(s)$ + _____ $CO(g)$
3. _____ $NH_4NO_3(s)$ → _____ $N_2(g)$ + _____ $O_2(g)$ + _____ $H_2O(g)$

In addition to balancing equations, varied practice with this concept in chemistry could involve other tasks, such as completing word problems like the following. You also could include them in the previous problem sets as *easy*, *medium*, or *challenge*.

> Although they often appear bright and shiny, aluminum objects are covered with a tight, invisible coating of aluminum oxide, (Al_2O_3) which forms when freshly-exposed aluminum (Al) reacts with oxygen. Write the balanced equation for this reaction.

Catch and Fix the Error

Another effective strategy in varied practice is to catch and fix the error. Similar to correcting false statements on a true-false quiz, this strategy encourages students to approach problems in a different or opposite way. They look at a proposed solution to a problem, find the error in that solution, and then correct the error. This can be a highly engaging experience applicable in numerous contexts. For example, if you give students the following balanced equations, students can find the errors, cross out the incorrect information, and correct the errors. In these equations, the types of errors are varied; any errors should be based on mistakes frequently made by students. With balancing equations, common issues arise with writing formulas correctly (and predicting products), providing the correct coefficient, reducing coefficients to the lowest common denominator, and so on.

1. ___3___ $CaOH_2$ + ___2___ H_3PO_4 → _____ $Ca_3(PO_4)_2$ + ___6___ H_2O

   Error: Parentheses should be placed around the OH in $CaOH_2$ so that it reads $Ca(OH)_2$.

2. ___2___ C_6H_{14} + ___19___ O_2 → ___6___ CO_2 + ___7___ H_2O

   Error: The equation is balanced incorrectly. CO_2 should have a coefficient of 12. H_2O should have a coefficient of 14.

3. ___2___ $Al_2(CO_3)_3$ + ___4___ H_3PO_4 → ___4___ $AlPO_4$ + ___6___ CO_2 + ___6___ H_2O

   Error: The equation is actually balanced correctly, but the coefficients should always be reduced to the smallest whole number ratio. Coefficients (in order) should be 1, 2, 2, 3, 3.

Construct Explanations

Students can also engage in NGSS science practices, such as constructing explanations, to vary the practice they have had in relation to balancing equations or other mathematics-related tasks. In the process of constructing their explanations, students must incorporate evidence to support any claim that they make. The evidence they'll use will likely come from past practice, and this serves to both encourage them to reflect on past learning and show readiness for new concepts. For example, you could give students writing prompts such as:

- Explain why the principle of conservation of mass necessitates that we write chemical equations in a balanced format.
- Create a list of four or five steps that describes the process of balancing equations. In your explanation, use a sample chemical equation for illustrative purposes. Be clear and thorough in your example.

Element 10: Examining Similarities and Differences

This element involves taking a critical look at the similarities and differences between two items related to procedural or declarative knowledge, which deepens students' understanding. For example, students can compare and contrast all manner of living organisms when making observations or classifying them into categories.

Element 10 focuses on the inherent value of discrimination in the scientific process. To categorize any sort of phenomenon, it is critical to understand similarities and differences. A great deal of scientific knowledge is organized in pairs: *mitosis* and *meiosis*, *particles* and *waves*, *prokaryote* and *eukaryote*, *endothermic* and *exothermic*, to name a few. When categorizing in pairs, similarities and differences are essential to properly group concepts and ideas. Similarities and differences are also foundational to tools, such as dichotomous keys, that are often used to classify plants and animals. In this section, we discuss specific strategies to build student capacity to discriminate effectively. Figure 4.3 shows the self-rating scale for this element.

Score	Description
4: Innovating	I engage in all behaviors at the Applying level. In addition, I identify those students who are not gaining new and deeper insights into the content they compare and design alternate activities and strategies to meet their specific needs.
3: Applying	Along with adequate guidance and support, I engage students in activities that help them identify similarities and differences in content and monitor students to ensure that they gain new and deeper insights into the content they compare.
2: Developing	I engage students in activities that help them identify similarities and differences in content and provide adequate guidance and support.
1: Beginning	I engage students in activities that help them identify similarities and differences in content but do not provide adequate guidance and support, such as ensuring that students identify important attributes on which to compare content and ensuring that students accurately identify the extent to which elements possess the identified attributes.
0: Not Using	I do not engage students in activities that help them identify similarities and differences in content.

Figure 4.3: Self-rating scale for element 10—Examining similarities and differences.

Teachers can implement the following strategies to create learning opportunities for students with regard to this element. After students engage in any activity using one of these strategies, teachers emphasize the similarities and differences between the topics as the basis for the exercise.

- Venn diagrams and constructed-response comparisons
- Dichotomous keys
- Sorting, matching, and categorizing

Venn Diagrams and Constructed-Response Comparisons

Venn diagrams, diagrams comprised of two overlapping circles, are powerful tools to help students organize and represent their comparative thinking. Each overlapping circle represents a different item, organism, or idea. The overlapping area represents the similarities that exist between the two items being compared. The areas where the circles don't intersect are reserved for unique characteristics or traits of each item being compared. You can use Venn diagrams in conjunction with a variety of other comparison strategies, such as constructed-response comparisons.

For example, in an elementary science lesson, students could compare any two organisms, such as a frog and a bird or a flower and a tree. Ask students to think about the many ways any two organisms are similar to one another. You might provide prompts to promote student thinking, such as the following.

- How does each organism get its energy (or what do they eat)?
- Where does the organism live?
- How does the organism move?
- How does each organism survive during periods of extreme weather?
- How does each organism reproduce and develop?

Students can begin organizing their thinking using the Venn diagram, with similar concepts recorded in the overlapping circles and unique characteristics recorded in the nonoverlapping circles.

Once students complete their Venn diagrams, they can use a strategy such as constructed-response comparisons to explain and expand on what is in the Venn diagram. These writing tasks can begin with a prompt such as, "How is a *frog* similar to and different from a *bird*?" In these paragraph-length responses, students synthesize what they found in their learning, highlighting similarities and differences and explaining how they are connected. In doing so, students must decide which similarities and differences they will include in their response and how to best frame their analysis to completely answer the question.

Dichotomous Keys

A *dichotomous key* consists of a series of statements with two choices in each step of the series. At each step, the user selects one choice. At the end of the series, if all choices have been correctly selected, the key provides the identity of a particular organism. Dichotomous keys are often used in biology to organize and identify closely related organisms. The essence of the dichotomous key is that the user receives two possible options for a particular trait; the choice then leads down a pathway until an accurate identification can be made. Science teachers can use these keys when studying insects, fish, trees, wildflowers, or other organisms (or items) that need to be classified. In elementary, middle, and high school science classrooms, the primary benefit of these tools is to use close inspection to understand the many ways in which the organisms are both similar and different. It can be helpful and efficient to model the use of the keys in the classroom setting with a specific set of identified samples. This allows the teacher to describe the nuances students should be looking for and answer any logistical questions about how to use the keys.

Once students grasp how to use dichotomous keys, they can engage in an enriching experience of putting this skill into practice, often in an outdoor experience. Students could use an insect or tree key to walk around the school property and identify organisms they find. They might then catalog the species they encounter to determine the species of particular specimens they are asked to identify or simply observe the various traits seen in the organisms under investigation. With cell phones, students can also take photos (particularly of insects), which allows them even more sustained and up-close investigation. Once students identify the samples they observe, they can check with the teacher, confirm with peers, or even use one of the many available phone apps to quickly identify a species as you take photos (truly remarkable!).

Finally, once students have some experience using a dichotomous key, they can create their own keys. For example, students could make a dichotomous key for what they wear. They can begin with two very broad categories and then continually narrow the choices down until they can sort all shoes into individual "shoe species." One of the most powerful and engaging aspects of this process is when students realize their key has inadvertently excluded or not accounted for some shoes students may be wearing. The iterative process of creating an initial model, refining that model in the face of the model's limitations, and then retesting the model makes students realize that it's OK if their first attempt isn't perfect; this is something scientists nearly always encounter in their work. This productive struggle is the process of high-effort learning that develops grit and creative problem-solving capabilities (Blackburn, 2018).

Sorting, Matching, and Categorizing

Science is full of opportunities to sort, match, and categorize. This is a powerful way for students to think critically about the similarities and differences between items they are studying and discover the rationale behind classification systems scientists have developed.

One of the most frequently used tools in middle school and high school science is the periodic table of elements. This table is constructed with periods (rows) that show recurring patterns moving from left to right across the table. Students can engage in many experiences to compare and contrast the characteristics used to develop the table, including electronegativity, ionization energy, atomic radius, melting point, and so on. These characteristics, once understood, can be powerful predictors of each element's properties, how each element reacts in a particular situation, and even facilitate prediction of the characteristics of elements yet to be discovered. Once students understand how the table is constructed, you can provide them with a set of data that describes a particular element, and students can determine in which family (column) and period (row) of the periodic table it belongs.

Another place science students encounter sorting, matching, and categorizing is in the study of rocks. Typically, students learn a great many characteristics about igneous, sedimentary, and metamorphic rocks, as well as the changes that occur over time to form the rock cycle. You might ask students to learn the names of specific rocks; this experience provides familiarity with some common rocks that students might experience in their daily lives or even their travels. However, an even more significant learning experience is for students to see the characteristics present in a rock sample, use those characteristics to place the rock in a category, and then construct the story of this rock.

Some students can ascertain these characteristics by visual inspection, while others rely on chemical tests such as reactivity with acid. For example, in the study of igneous rocks, students can visually inspect a sample to tell if the rock cooled inside the earth or at or near the surface of the earth. This also gives clues as to whether the rock cooled slowly or quickly, trapped gas in the process, and so on. In addition, based on the color of the rock (and the minerals in the rock), students should be able to tell whether this rock formed on an oceanic plate, a continental plate, or at the boundary of these two types of plates. When students learn skills for categorizing rocks, they are then able to apply this to any rock sample they encounter, teaching them patterns of thinking that move beyond the samples studied in class.

Element 11: Examining Errors in Reasoning

Element 11 examines how students approach errors in scientific thinking and experimentation. Science proceeds only through consistently surveying the available evidence and self-correcting when theories are no longer supported. Some of the greatest scientific discoveries in human history have been the result of errors or serendipity in the lab. Teachers can certainly play a role in identifying and correcting errors, but it is extremely powerful when students are trained to recognize errors and make their own corrections. When students thoughtfully examine their own thinking or the information teachers present to them, they deepen their understanding of science.

This section presents strategies to engage students in the identification and correction of scientific errors. Figure 4.4 presents the self-rating scale for this element so teachers are better equipped to determine how they can support students to examine potential errors in reasoning.

Score	Description
4: Innovating	I engage in all behaviors at the Applying level. In addition, I identify those students who are not recognizing errors in their own thinking or that of others and taking steps to rectify those errors and design alternate activities and strategies to meet students' specific needs.
3: Applying	Along with adequate guidance and support, I engage students in activities that help them examine errors in reasoning and monitor students to ensure that they recognize errors in their own thinking or that of others and take steps to rectify those errors.
2: Developing	I engage students in activities that help them examine errors in reasoning and provide adequate guidance and support.
1: Beginning	I engage students in activities that help them examine errors in reasoning but do not provide adequate guidance and support, such as explicitly teaching common types of errors in reasoning and providing practice in identifying such errors.
0: Not Using	I do not engage students in activities that help them examine errors in reasoning.

Figure 4.4: Self-rating scale for element 11—Examining errors in reasoning.

The following strategies present options for teachers to aid students in examining their own or others' errors in reasoning.

- Identifying errors of misinformation
- Finding errors in the media

Identifying Errors of Misinformation

With the introduction of NGSS, science teachers more frequently ask students to identify errors of misinformation (through argumentation) and construct scientific explanations (NGSS Lead States, 2013). Each of these closely-related science practices asks students to develop a claim, provide evidence, and justify the claim with scientific reasoning. Science teachers use a variety of approaches to improve students' confidence and competence in this area. While generating these writing samples is a key skill for students to develop, scientists must also be able to identify errors of misinformation.

A valuable strategy in developing students' skill in writing their own arguments (or between a first and final draft) is to have students analyze sample arguments or explanations. In doing so, students are able to detect errors in others' writing or reasoning used to support the claim and evidence. By doing so, they avoid the same mistakes made in the work they are reviewing and become more familiar with the criteria used to identify proficient work. Students can then confidently construct their final work. Table 4.1 outlines various strategies for examining errors in reasoning.

Table 4.1: Examining Errors in Reasoning

Strategy	Description
Identifying errors in faulty logic	The teacher asks students to find and analyze errors of faulty logic. Errors of faulty logic refer to situations in which sound reasons do not support a conclusion. Specific types of errors in this category include contradiction, accident, false cause, composition, division, begging the question, evading the issue, and arguing from ignorance.
Identifying errors of attack	The teacher asks students to find and analyze errors of attack. Errors of attack happen when a person focuses on the context of an argument rather than the argument itself when trying to refute the other side.
Identifying errors of weak reference	The teacher asks the students to find and analyze errors of weak reference. Specific types of these errors include using sources that reflect biases, using sources that lack credibility, appealing to authority, appealing to the people, and appealing to emotion.
Identifying errors of misinformation	The teacher asks students to find and analyze errors of misinformation. Two types of misinformation errors are confusing the facts and misapplying a concept or generalization.
Practicing identifying errors in logic	The teacher uses practice exercises to help students identify errors in logic. Typically, these exercises describe a scenario in a few sentences and ask students to identify the reasoning error present in the scenario. Students might select the answer in a multiple-choice or matching format or be asked to recall the answer from memory.
Finding errors in the media	The teacher provides students with footage from political debates, televised interviews, commercials, advertisements, newspaper articles, blogs, and other sources and asks them to find and analyze errors in reasoning that underlie the messages therein.
Examining support for claims	The teacher asks students to examine the support provided for a claim by analyzing the grounds, backing, and qualifiers that support it. Grounds are the reasons given to support a claim, and backing is the evidences, facts, or data that support the grounds, while qualifiers address exceptions or objections to the claim.
Judging reasoning and evidence in an author's work	The teacher asks students to apply their knowledge of reasoning and argumentation to delineate and evaluate the arguments present in a text. Students read a text and identify the claim, grounds, backing, and qualifiers. Students must decide whether the reasoning is valid or logical (containing no errors) and whether the supporting evidence is sufficient and relevant.
Identifying statistical limitations	The teacher asks students to find and analyze errors that commonly occur when using statistical data to support a claim. The major types of statistical limitations for students to be aware of are: (1) regression toward the mean, (2) conjunction, (3) base rates, (4) limits of extrapolation, and (5) the cumulative nature of probabilistic events.
Using student-friendly prompts	The teacher phrases prompts and questions in nontechnical language to trigger students to look for certain types of errors; for example, asking students to look for arguments "getting off topic" rather than "evading the issue."
Anticipating student errors	The teacher identifies errors that students are likely to make during a lesson. When presenting content, the teacher alerts students to the potential problems. For example, when a teacher introduces the process for finding density, he or she reminds students that the density remains constant no matter the volume, provided the substance stays the same.

continued →

Strategy	Description
Avoiding unproductive habits of mind	Unproductive habits of mind are those that hinder us from completing complex tasks. To counteract unproductive habits, the teacher reinforces the following productive habits of mind: staying focused when answers and solutions are not immediately apparent, pushing the limits of your knowledge and skills, generating and pursuing your own standards of excellence, seeking incremental steps, accuracy, and clarity, resisting impulsivity, and seeking cohesion and coherence.

Source: Marzano, 2017, pp. 41–42.

Finding Errors in the Media

Unfortunately, the media often reports science stories that include errors. Such errors can include overstated political rhetoric, inaccurate reports on community issues, or even intentionally inaccurate portrayals of products in commercials. Movies that include science concepts typically hire scientists to review the movie to point out inaccuracies and inconsistencies (Smaglik, 2014). Students can do the same things with everyday media. (Refer to table 4.1, page 63, for additional ways to combat errors in reasoning.)

To use this strategy, a teacher might collate a portfolio of media files with known scientific errors that relate to specific topics within the curriculum. After reviewing the files, a teacher can present students with a TV commercial, YouTube video, or form of media and ask them to find the science mistakes. In addition to video material, errors in blogs, newspapers, and other print sources can provide learning experiences in which students must discriminate sound science from pseudoscience and falsehoods. If these types of activities are consistently instituted throughout the school year, students will begin looking and listening more critically, becoming more prudent consumers of media sources. In addition, a teacher might invite students to bring in samples of items that have errors (as well as those that support class topics of learning). Some examples include vague, unsubstantiated health claims on vitamin packaging or faulty physics content in comic books.

GUIDING QUESTIONS FOR CURRICULUM DESIGN

This design question focuses on practicing and deepening lessons: *After presenting content, how will I design and deliver lessons that help students deepen their understanding and develop fluency in skills and processes?* The following questions, which are aligned to each of the elements in this chapter, can guide teachers to plan instruction. Each represents distinct activities rather than a sequential order when planning. In other words, if the content the teacher presents is procedural (skills, strategies, and processes), structured practice is necessary. If it's more declarative in nature (terminology, facts, generalizations, and principles), teachers use strategies in which students examine similarities and differences.

- **Element 9:** How will I help students engage in structured practice?

_____

_____

_____

_____

_____

- **Element 10:** How will I help students examine similarities and differences?

_____

_____

_____

_____

_____

_____

_____

- **Element 11:** How will I help students examine errors in reasoning?

_____

_____

_____

_____

_____

_____

_____

Summary

Science teachers plan lessons to help students practice and deepen both procedural knowledge—skills, strategies, and processes—and declarative knowledge, which involves terminology, facts, generalization, principles, or concepts. With procedural knowledge and skills, such as the science and engineering practices, teachers select instructional strategies that focus on structured practice. For learning that is more declarative, such as with the disciplinary core ideas, teachers can design lessons that examine similarities and differences. When choosing strategies that examine errors in reasoning, activities can be geared toward both types of knowledge. In the process, teachers should continually be looking to create opportunities for students to apply what they learn.

CHAPTER 5

Conducting Knowledge Application Lessons

With the introduction of NGSS, science teachers are reminded of the value of ensuring that knowledge (disciplinary core ideas, or DCIs) are not standalone bits of information (NGSS Lead States, 2013). Rather, students need to explore and apply these core ideas within the context of science practices and crosscutting concepts. This application and transfer of science ideas in multiple ways deepens their understanding and connection of these ideas. Once students experience the content from direct instruction lessons and engage in practicing and deepening lessons, they are ready to apply newly acquired information and skills. During knowledge application lessons, teachers require students to use what they have learned in unique situations. This means students must go beyond what they have learned in class and generate new awareness about core ideas. The following elements are important to knowledge application lessons.

- **Element 12:** Engaging students in cognitively complex tasks
- **Element 13:** Providing resources and guidance
- **Element 14:** Generating and defending claims

Teachers should use these elements in conjunction with one another. That is, they should typically employ all three when assigning knowledge application tasks.

Element 12: Engaging Students in Cognitively Complex Tasks

Cognitively complex tasks are tasks that require students to use the content teachers have addressed in class in unique ways. These tasks typically require students to transfer scientific content knowledge to both produce and support claims. Figure 5.1 (page 68) contains the self-rating scale for this element. The lessons in this category are at the heart of quality science instruction as students create predictions, investigate scenarios, and try to match their observations with their predictions. Using this information, students create a conclusion that explains their findings and provides connections to their understanding of scientific principles.

Score	Description
4: Innovating	I engage in all behaviors at the Applying level. In addition, I identify those students who are not accurately carrying out the important parts of those tasks and design alternate activities and strategies to meet their specific needs.
3: Applying	Along with adequate guidance and support, I engage students in activities that help them apply their knowledge through cognitively complex tasks and monitor students to ensure that they accurately carry out the important parts of those tasks.
2: Developing	I engage students in activities that help them apply their knowledge through cognitively complex tasks and provide adequate guidance and support.
1: Beginning	I engage students in activities that help them apply their knowledge through cognitively complex tasks but do not provide adequate guidance and support, such as clearly articulating the steps in the complex task and modeling the steps to the task.
0: Not Using	I do not engage students in activities that help them apply knowledge through cognitively complex tasks.

Figure 5.1: Self-rating scale for element 12—Engaging students in cognitively complex tasks.

The strategies associated with engaging students in cognitively complex tasks require several mental steps for students as they think deeply about newly learned content and apply it in novel situations.

- Experimental-inquiry tasks
- Invention tasks

Experimental-Inquiry Tasks

Experimental-inquiry tasks begin with student observation of a phenomenon, which leads students to ask questions or answer a testable question. A good example of an experimental-inquiry task is the classic "water drops on a penny" experiment. The teacher shows students a penny and then adds a drop of water to the top of the penny from an eye dropper. Then, the teacher poses the following question: "How many drops of water can you fit on the head of a penny?" This question and subsequent experimentation are valuable experiences for elementary students. First, ask students to predict how many drops of water will fit on the head of a penny before the water runs over the edge. Each student gets a penny, a water dropper, and some water. Students then place water drops, one at a time, onto the head of the penny. The penny will hold many more drops of water than most students predict. This difference in number of water drops creates dissonance for students and pushes them to ask, "Why?" Students and teachers work together to create explanations for their observations.

Quality examples of experimental inquiry point students to a deeper understanding of a scientific phenomenon. Experimental inquiry leads students to ask questions, pose hypotheses, conduct experiments, address misconceptions, and revise their understanding based on the learning acquired throughout the process. It is important for the teacher to continually refocus students on scientific understanding instead of the experimental procedure. If students become too process-oriented or procedure-focused, then they're simply following a predetermined recipe and aren't engaging with the cognitive complexities of the scientific problem or phenomenon.

To avoid this trap, teachers must give students only the minimal starting information necessary and coach them through a process of discovery. Teachers must plan for faulty student hypotheses and ill-considered conclusions. Most importantly, the teacher must direct students to incorporate the scientific content into their conclusions. In the case of the penny, it is important for students to understand that water drops are attracted to one another and to the penny in a way that allows for the addition of multiple of drops (typically fifteen to twenty-five). Students should leave this experiment with a better understanding of surface tension, water adhesion, and water cohesion.

Invention Tasks

Giving students the opportunity to demonstrate their understanding of scientific laws often comes in the form of a practical or hypothetical student invention. When teachers ask students to invent something, students must consider the problem, conduct research, engage in a design process, and test the design (either in the lab or in a theoretical scenario). All these tasks require students to address the scientific law properly or the design simply will not work when put to the test. In the classroom, teachers can simply ask students to create a product that solves a problem. A common example is the egg drop. For this activity, students have to design a vessel that will keep an egg safe when dropped from a significant (greater than three meters) height. They must demonstrate their understanding of forces and impulse when creating the vessel.

One challenge in implementing an invention lesson is how to determine successful learning. The success or failure of the invention (egg survives or cracks) should not be the fundamental measure of student success. Teachers should design a few questions to understand the extent of the learning students mastered. A student might design a vessel that allows the egg to break, but complete and valid explanations for these questions could represent quality work on this assignment. These conclusion questions should challenge the student to place the invention into scientific context that demonstrates high-level (cognitively complex) thinking, such as the following.

- Why did the invention succeed or fail?
- What have you learned about forces and impulse?
- How would you improve this invention in the future?

Answering these conclusion questions provides a nice summary of student thinking at the end of the egg-drop experiment. Students may orally answer these questions in front of the class (or in small groups) as they hold their vessels and explain. In addition, these questions may be formalized as a written, scored assessment. In either case, using conclusion questions is an effective strategy for getting a snapshot of student thinking that can inform future instruction. They also can serve as reflections for students that document takeaways from the learning experience.

Element 13: Providing Resources and Guidance

Engaging in cognitively complex tasks is challenging for many students. Consequently, teachers must be available and prepared to provide adequate resources and guidance. The science classroom should provide a wealth of resources and investigative tools that can support creative discovery. In elementary school, this may involve safe items like paper, scissors, tape, pots, dirt, and seeds. In the middle and high school levels, experimental toys may include scientific equipment like balances, Bunsen burners, and chemical mixtures in burettes. Regardless, true scientific inquiry in the classroom relies on a sense of playfulness and exploration. Therefore, it is important to make sure students have choice and the proper tools to play and experiment with different possibilities. In distinct opposition to the idea of free play, it is the teacher's responsibility to ensure that the play is safe, guided, and tied to specific scientific learning outcomes. Students also should take ownership in the process and take on more responsibility as they gain facility and knowledge.

Figure 5.2 (page 70) shows the self-rating scale for this element.

Score	Description
4: Innovating	I engage in all behaviors at the Applying level. In addition, I identify those students who are not gradually taking over responsibility for the completion of the complex tasks and design alternate activities and strategies to meet their specific needs.
3: Applying	I engage in activities to provide resources and guidance to students as they engage in cognitively complex tasks without significant errors or omissions and monitor the extent to which students are gradually able to take over responsibility for the completion of the tasks.
2: Developing	I engage in activities to provide resources and guidance to students as they engage in cognitively complex tasks without significant errors or omissions.
1: Beginning	I engage in activities to provide resources and guidance to students as they engage in cognitively complex tasks but do so with errors or omissions, such as not being readily available to students and not having critical resources available to students as they work on their complex tasks.
0: Not Using	I do not engage in activities to provide resources and guidance to students as they engage in cognitively complex tasks.

Figure 5.2: Self-rating scale for element 13—Providing resources and guidance.

This section addresses the following strategies for providing resources and guidance.

- Providing resources
- Circulating around the room
- Offering feedback
- Creating cognitive dissonance

Providing Resources

Giving students real-world data (or access to data on the internet) they can use to further extend their understanding is very helpful as they participate in science lessons and activities. When using real-world data, students receive an implicit message that the work they are doing is meaningful and relevant. Made-up scenarios and fake data do not have the same power to engage. In addition, real-world data can be connected to actual problems that students may be aware of in their community or in the world at large. Teachers can source useful qualitative and quantitative data from videos, podcasts, articles, print books, and website graphics (to name a few). Although teachers can generally construct theoretical data sets more quickly than finding primary source data, it is worth the extra effort because of the all the possible learning extensions.

For example, if students are conducting research or experiments related to weather, a teacher-provided resource could include access to average temperatures, wind speed, or humidity from the National Weather Service website (www.weather.gov). Consider how a middle school teacher might want students to use the website to track daily weather data. Their end goal is to collectively use the data to develop a better understanding of how weather patterns move across the United States and the planet. Assign each student a different U.S. city. At the end of the week, students can share their information, find similarities and differences in the data, and try to explain how weather patterns change across the country.

Circulating Around the Room

Circulating around the room is a time-honored strategy that helps teachers monitor group work and address individual student learning needs. This proactive approach decreases student misbehavior and increases student engagement (Haydon, Hunter, & Scott, 2019). During an experiment or inquiry-based activity, teachers often move around the room and assist students in their work. The questions teachers get from students are almost always about procedures or how to operate lab equipment. In these situations, teachers should develop two questions in advance that explore the underlying scientific content.

For example, many elementary schools use an inquiry activity related to flubber (a non-Newtonian fluid made from cornstarch). Flubber flows like a liquid under low stress but becomes firmer and breaks like a solid at high stress. Questions related to this activity might include: "Is flubber a liquid or a solid?" or "What makes flubber change in the way it behaves?" These questions require students to think about the science, not the procedure. Stop at each table and ask these questions as you move around the room to refocus students' attention on better understanding the science behind the experiment. These questions are key to students making connections between the task at hand and the scientific phenomena being explored.

Offering Feedback

Jan Chappuis (2015) describes five essential characteristics of effective feedback to help support and improve student learning.

1. **Outcome based:** Points attention to the intended learning, identifying strengths and offering specific information to guide improvement
2. **Timely:** Takes place during the learning process, while there is time to act on it
3. **Complete:** Addresses partial understanding
4. **Learning, not telling:** Does not do the thinking for students
5. **Student focused:** Limits corrective information to what students can actually act on

In science classes, teachers often provide feedback to assessments that simply point out mistakes: red checkmark, incorrect answer, or just plain wrong. It is important to consider creating a framework for providing students with true growth-based feedback when grading assignments. Student feedback should follow Chappuis's recommendations.

Creating Cognitive Dissonance

Cognitive dissonance is the mental discomfort that an individual experiences while trying to hold multiple conflicting beliefs, theories, or ideas. Research hypothesizes that this period is often brief because the inconsistencies compel the individual to act to resolve his or her mental discomfort (Harmon-Jones, Harmon-Jones, & Levy, 2015). An effective tool for creating cognitive dissonance in the science classroom is a class-wide demonstration. When the teacher performs a demonstration before students understand the science behind it, the question of "Why did this happen?" can drive interest and instruction. Students will want to understand the phenomena behind the demonstration. Provide students with a few guiding questions to help them collect observations.

- "What did you see?"
- "What did you hear?"
- "What was interesting about the observation?"

You can carry out the demonstration while students observe and record. A great example of this technique is demonstrating how a static charge can influence a stream of water. Allow a small stream of water to fall from the faucet. Then charge up a balloon or plastic rod with static electricity. When you place the charged item near the stream of water, the water will bend toward the charge. Ask students to record everything they see and share observations with their partners. Ask students if all liquids would bend toward the charge.

This demonstration initially puts students in a state of discomfort because they've most likely never seen water bend before. They probably believe water is a neutral substance unaffected by manipulations like those in the demonstration. After seeing the demonstration, they come to the realization that water must have hidden properties that allow it to interact with electricity. In this manner, the teacher is able to engage students in learning about water molecule structure and its partial polarity.

Element 14: Generating and Defending Claims

The ultimate goal of cognitively complex tasks is to grow students' ability to generate new conclusions and defend their claims by providing evidence. This goal is imperative because students can transfer this skill to a wide variety of scientific tasks once it is realized. In addition, once students become skilled in this process, teachers may engage in more inquiry- and problem-based scenarios in which students can construct knowledge for themselves and take ownership in the learning process. This is especially important when students interact with science and engineering practices (SEPs) such as constructing explanations; engaging in argument from evidence; and obtaining, evaluating, and communicating information.

Figure 5.3 shows the self-rating scale for this element.

Score	Description
4: Innovating	I engage in all behaviors at the Applying level. In addition, I identify those students who are not developing logical arguments regarding their claims and design alternate activities and strategies to meet their specific needs.
3: Applying	Along with adequate guidance and support, I engage students in activities that help them generate and defend claims and monitor students to ensure that they have developed logical arguments regarding their claims.
2: Developing	I engage students in activities that help them generate and defend claims and provide adequate guidance and support.
1: Beginning	I engage students in activities that help them generate and defend claims but do not provide adequate guidance and support, such as providing a clear model of the nature of an effective argument with its related parts and providing adequate practices in analyzing and constructing arguments.
0: Not Using	I do not engage students in activities that help them generate and defend claims.

Figure 5.3: Self-rating scale for element 14—Generating and defending claims.

The strategies associated with this element focus on components of a well-structured and developed argument.

- Providing grounds
- Formally presenting claims

Providing Grounds

After students make a claim, they need to provide supporting evidence or grounds to explain and support their claim. They can do this by adding *because* to the end of the claim. Again, this is directly in line with science practices such as constructing explanations or engaging in argument from evidence.

This is a great opportunity to expose misunderstandings in students through a before-and-after activity. Begin by providing a prompt that students might not completely understand, such as "It is warmer in the summer than in the winter because _____." Many students might suggest that the sun is closer to Earth. After providing new information that outlines that temperature variations are really due to the tilt of

Earth compared to the sun, students can return to the original question and provide grounds by answering "because _____." The power of this activity is that it requires students to confront their misconceptions and replace them with the new information presented in class (see figure 5.4).

Before
It is warmer in the summer than in the winter because . . .
How do you know this is true?

After
It is warmer in the summer than in the winter because . . .
How do you know this is true?
How has your thinking changed today?

Figure 5.4: Before-and-after activity example.

Other examples might include the following.

- The wind mostly blows from the west because . . .
- Flowers won't grow on the north side of my house because. . . . Would the same be true in Australia?
- Tap water is not pure because . . .
- Wolves grow to be larger than frogs because . . .

When students provide supporting evidence or grounds, they are providing a scientific rationale for their claim. The grounds provide an easy way for the teacher to identify if students have a working understanding or if they are still harboring a misconception.

Formally Presenting Claims

To truly understand science, students need to understand the scientific review process. As scientists develop new ideas and understandings, they must present those ideas to other scientists for peer review. This process can be a powerful instructional tool in a science classroom. Many students believe that all claims are equally valid because they've heard that they must respect the opinions of others. In addition, some students have read about new experiments that make exciting claims about the efficacy of drugs for particular diseases. Unfortunately, many students believe if something is in print or on the internet, it must be true. In science, claims are held to a higher standard. Students should understand that scientists need to present their ideas, support them with evidence, and be prepared to defend their conclusions to competing scientists for the scientific community to hold the ideas in any regard.

A good example of this in practice is asking students to determine the factors that affect the period of a pendulum. Assign small student groups a different factor, length of the string, mass of the pendulum bob, and release angle of the pendulum. Each group must experiment with its variable or factor (keeping the other variables constant) and collect meaningful data. Using this information, group members develop a claim related to whether their variable changes the period of the pendulum. Having groups present their claims and evidence to the class at the end of the experiment enriches the learning and deepens student understanding.

After the presentations, the class as a whole can develop a unified theory of pendulums. This experiment represents powerful learning for students because their initial thoughts on what impacts the period of a pendulum are often wrong, forcing them to rely on their experimental data to create their claims. Experiences such as this can lead to students engaging in the science practice of constructing scientific explanations.

GUIDING QUESTIONS FOR CURRICULUM DESIGN

To implement knowledge application lessons, this is the design question: *After presenting new content, how will I design and deliver lessons that help students generate and defend claims through knowledge application?* Consider the following questions aligned to the elements in this chapter to guide your planning.

- **Element 12:** How will I engage students in cognitively complex tasks?

_____

_____

_____

_____

_____

_____

_____

- **Element 13:** How will I provide resources and guidance?

_____

_____

_____

_____

_____

_____

_____

- **Element 14:** How will I help students generate and defend claims?

_____

_____

_____

_____

_____

_____

_____

Summary

After leading lessons to introduce new science content, teachers must also design experiences for students to apply what they learned in a novel situation so they can demonstrate mastery, both with science concepts and a variety of science and engineering practices. The ultimate goal of cognitively complex tasks, a hallmark of knowledge application lessons, involves students generating their own claims and providing evidence for their conclusions. Teachers can employ a host of instructional strategies within all types of lessons—direct instruction, practicing and deepening, and knowledge application—which is the topic of the next chapter.

Using Strategies That Appear in All Types of Lessons

Chapters 3, 4, and 5 represent specific types of lessons, each with unique purposes and strategies. Chapter 3 deals with direct instruction lessons, chapter 4 with practicing and deepening lessons, and chapter 5 with knowledge application lessons. The instructional strategies we discuss in those chapters would probably appear in the context of their respective lessons. For example, teachers use chunking primarily when introducing new content, structured practice sessions primarily when developing fluency in a procedure, and cognitively complex tasks when having students apply their knowledge. It is important to note that there is no particular order to these chapters or types of lessons. Students should begin applying their knowledge (chapter 5) well before they have completely deepened or mastered their understanding. Through this practice, students gain a fuller understanding of what they know and don't know.

In addition to strategies from chapters 3–5, teachers can and should use many additional strategies within all three types of lessons discussed in those chapters. These strategies help students continually integrate new science knowledge with prior knowledge, make connections through the crosscutting concepts, demonstrate understanding through the science and engineering practices, and revise their understanding of the content accordingly. A key element of this chapter's strategies is that they ask students to make connections between what they have learned (transfer) and put those learnings into practice. The following elements and the strategies embedded within them in this chapter help students perform these functions.

- **Element 15:** Previewing strategies
- **Element 16:** Highlighting critical information
- **Element 17:** Reviewing content
- **Element 18:** Revising knowledge
- **Element 19:** Reflecting on learning
- **Element 20:** Assigning purposeful homework
- **Element 21:** Elaborating on information
- **Element 22:** Organizing students to interact

As mentioned previously, to see all strategies within every element, refer to appendix A (page 149).

Element 15: Previewing Strategies

Previewing helps students activate prior knowledge so they make connections with new knowledge. The act of previewing requires students to bring past learning into the present and anticipate how the information or skills might apply to the new lesson or scenario. In addition, these strategies provide opportunities for students to share knowledge from their personal backgrounds and lived experiences. Previewing can serve to pique student interest and establish a baseline of entry knowledge.

Figure 6.1 shows the self-rating scale for previewing strategies.

Score	Description
4: Innovating	I engage in all behaviors at the Applying level. In addition, I identify those students who are not making connections between new content and what they have learned previously and design alternate activities and strategies to meet their specific needs.
3: Applying	Along with adequate guidance and support, I engage students in activities that help them preview upcoming content and monitor students to ensure that they are making connections between new content and what they have learned previously.
2: Developing	I engage students in activities that help them preview upcoming content and provide adequate guidance and support.
1: Beginning	I engage students in activities that help them preview upcoming content but do not provide adequate guidance and support, such as demonstrating the purpose for previewing and providing adequate time for students to preview content.
0: Not Using	I do not engage students in activities that help them preview upcoming content.

Figure 6.1: Self-rating scale for element 15—Previewing strategies.

We associate the following strategies with this element.

- Informational hooks
- What do you think you know?
- Preview questions

Informational Hooks

One of the most exciting parts of being a science teacher is using scientific phenomena to surprise, engage, and stimulate further inquiry. It is important for teachers to use these natural connections and hooks to draw students into learning science (and show them why science teachers love science so much!). These demonstrations of science phenomena amaze and excite students, paving the way for students to ask questions, design experiments, construct explanations, and carry out all kinds of science practices.

For example, students can explore the use of a pinhole camera to model the function of the eye. Not only is the pinhole camera a great tool if your students ever have the opportunity to witness a solar eclipse, but it also serves as a model of how the eye functions. Depending on students' age, teachers can make many connections. For elementary students, the pinhole camera could simply show the passage of light through an opening for display on the backside of the box. Older students can go into much greater depth. This could include calculating (or experimenting with) the aperture size since larger openings allow for a brighter image, but smaller openings provide a sharper image. All ages can learn how these tools simulate what a camera does to capture light as an image or for why the image appears inverted. There are numerous video explanations online for how pinhole cameras work to provide additional facts and information to support this activity.

Another similar and highly effective hook for students is the use of demonstrations. For example, when exploring gas laws and how gases behave under varying conditions, students can see the power of gases when the air is removed from a vacuum chamber. Effective samples to use for this activity include marshmallows, shaving cream, or a simple beaker of water. In each of these examples, very distinct visible changes occur due to the expansion of the gases in the samples when subjected to decreased surrounding air pressure. Marshmallows grow in volume; shaving cream expands; and water boils. Similarly, you can provide a powerful demonstration of gas laws with the crushed can demo (heating a small amount of water in a can before inverting and plunging it into a cold-water bath). Due to the sudden change in temperature, the gases inside the can move much slower, dropping the pressure inside the can. The air pressure outside the can is now much greater than inside, allowing normal air pressure of 14.7 psi to forcefully crush the can.

Students in physical science or chemistry are typically excited to see things catch on fire or explode. While safety considerations are paramount when doing demonstrations of any type, many such demonstrations can safely draw in students and prepare them for learning. For example, the electrolysis of water results in hydrogen and oxygen gas. Students are amazed and curious when they see this substance that puts out fires (water) broken down into a gas used for combustion and another that is explosive.

Finally, video clips can bring all sorts of hooks into the classroom. For example, historical events such as the Hindenburg explosion can serve as a hook into learning many science principles while also providing historical context, exploring the nature of science, or even discussing contemporary science applications. You can model this same science, with appropriate safety precautions, by exploding a hydrogen-filled balloon. This can demonstrate a wide range of principles, including an exothermic reaction. Students can easily feel the heat coming off the balloon after it explodes.

What Do You Think You Know?

For this strategy, ask students to individually write down what they think they know about a topic. Students usually enjoy sharing their knowledge and can often laugh at the obvious knowledge gaps. When students write down what they already know, this helps reinforce what they have already learned in a visual way. When students use whiteboards or huddleboards, group thinking is visible to all in the classroom and can be shared through brief presentations or gallery walks (students circulate around the room to view each other's boards).

You also may employ this strategy individually with students. Students write down their initial thoughts on paper and save them for a later part of the unit. Later in the unit, students retrieve their papers and use them to engage in discussions about how their thinking has changed, if they had initial misconceptions, and what they could add based on new learning.

As was clearly demonstrated in the popular TV show *MythBusters* (Savage, Hyneman, Wolkovitch, et al., 2003–2018), students often come into science with many misconceptions of how the world works. In these myth-busting situations, it takes several, repeated, varied, and intentional experiences to replace these misconceptions. These can come from ideas or thoughts students have heard from parents, read in a legitimate source, or even experienced (for example, their body being thrown back into the seat when a car accelerates). Other commonly held misconceptions include phases of the moon. Students might be curious about what causes moon phases and eclipses. For example, many students would provide the same explanation for a new moon as they would for a lunar eclipse. Another large-scale misconception might be what causes ocean tidal changes. By asking students what they think causes these phenomena, teachers can activate prior knowledge, push students to explain their thinking, and then plan instruction to guide their learning. You might refer to

Page Keeley's (n.d.) popular *Uncovering Student Ideas in Science Series*, which probes students' understanding of science. This series includes multiple volumes that explore a range of science concepts.

Along with asking students what they think they know (looking back), prediction is an engaging teaching strategy for looking forward, connecting what you think you know to what you think will happen. For example, teachers can have students make a prediction for what they think will happen in a lab investigation. This goes beyond simply making a hypothesis by making specific predictions about the ways variables will interact and the magnitude of the interactions. In high school courses, when students test various variable settings or levels, they can also predict if the relationship will be direct, inverse, or linear.

Making predictions about any future learning event is a powerful science strategy for several reasons. When making predictions, students naturally pay closer attention to the outcome, as their interest has been heightened. In addition, they are emotionally connected to the outcome, hoping to be right and letting others know they are right.

In his book, *Small Teaching*, James M. Lang (2021) points out another important reason that prediction is so effective. Predicting engages students to think deeply and seek connections that enable accurate predictions. When students make predictions about what might happen, they force themselves to "search around for any possible information you might have that could relate to the subject matter and help you make a plausible prediction" (Lang, 2021, p. 49). In this way, students enact prior knowledge and bring it into the scenario.

Preview Questions

Preview questions and preassessments can serve very similar functions to using informational hooks and having students write what they think they know. Preview questions serve to pique students' curiosity and activate prior knowledge. Similarly, preassessments expose students to a lesson's big ideas and get them thinking about new topics of study. In both cases, do not use these questions for grading but to drive student engagement and get an initial understanding of student thinking to guide effective instruction.

Preview questions can stimulate fun brainstorming and drive students to generate explanations and consider concepts such as cause and effect. Constructed effectively, they can often put a smile on students' faces or cause them to think about something in a way they had never imagined. For example, when studying various processes (such as digestion, photosynthesis, or cellular respiration), ask students, "Where does a tree get its mass?" Students will often offer ideas like the soil, the water, or fertilizer as the source of increasing mass. Similarly, you can ask students to consider all the food they eat compared to the waste they produce. Where did it all go? Both answers come in the form of gas exchange—trees get their mass by taking in carbon dioxide and then fixing it into carbohydrates, while most of the mass of the food we eat is eventually given off through exhaled carbon dioxide.

You might ask a similar type of question such as, "What would happen if you only ate peanut butter for one year?" This question could probe understandings related to nutrients as well as disorders and diseases that might occur when diet goes too far out of balance. You can explore these kinds of outlandish questions in the popular book *What If? Serious Scientific Answers to Absurd Hypothetical Questions* (Munroe, 2014).

Element 16: Highlighting Critical Information

Students often miss critical information unless they make specific efforts to highlight key concepts in memorable ways. Students often must hear something multiple times in many different contexts to assimilate key concepts. In addition, students shouldn't simply *hear* key concepts; teachers should present and repeat them using visual cues and other multisensory approaches. A teacher should never simply assume that repetition is leading to assimilation of the information; they should actively seek evidence of learning from students.

Figure 6.2 depicts the self-rating scale for highlighting critical information, which helps students focus on information that is critical to their learning.

Score	Description
4: Innovating	I engage in all behaviors at the Applying level. In addition, I identify those students who are not aware of critical information and how its pieces fit together and design alternate activities and strategies to meet their specific needs.
3: Applying	I engage in activities to highlight critical information for students without significant errors or omissions and monitor the extent to which students are aware of critical information and how its pieces fit together.
2: Developing	I engage in activities to highlight critical information for students without significant errors or omissions.
1: Beginning	I engage in activities to highlight critical information for students but do so with errors or omissions, such as highlighting information that is not critically important and not putting adequate emphasis on critical information.
0: Not Using	I do not engage in activities to highlight critical information for students.

Figure 6.2: Self-rating scale for element 16—Highlighting critical information.

The following strategies align to the element of highlighting critical information, which is the focus of this section.

- Repeating the most important content
- Using visual activities
- Using dramatic instruction to convey critical content

Repeating the Most Important Content

One of the most effective ways to ensure students prioritize key learnings is to repeat the most critical content or disciplinary core ideas (Brown et al., 2014; Lang, 2021). While there is much more to learning than just hearing ideas, continued reinforcement helps students identify the most important content and understand how key elements relate. Repetition also helps them form connections between concepts as the year moves along.

An effective way to do this is to create visual representations of key terms or concepts in the classroom. We have seen colleagues use whiteboards (or other writing surfaces) to keep a visual, multiday running storyline that wraps around the classroom. This allows teachers to continually reinforce and connect previous concepts. Since these are visible to all students, the teacher can reference the growing body of knowledge and ask students to add to or modify it. For example, at the start of a lesson or unit, a teacher can refer to prior learning and then state, "Today we are going to build on the concept of . . ." or "Today we are going to apply our earlier understanding of. . . ." This also gives students an opportunity to reflect and see the learning story, particularly during those quiet moments in class. This also reminds the teacher of the value of connecting formulas, concepts, and key ideas. Continued retrieval of prior knowledge, particularly when students need to make these connections, is directly in line with best practices in learning theory (Brown et al., 2014; Lang, 2021).

Using Visual Activities

Asking students to visualize their learning and create visual representations can help them cement concepts in their minds. Visual representations should be active, constructive, and interactive to best engage students and support effective learning (Ainsworth & Scheiter, 2021). We highlight two variations for doing this. First, when students are preparing for a lab, it is helpful to have them draw out a lab procedure. This typically involves drawing a sketch of the lab apparatus (for example, ring stand, Bunsen burner, glassware, probes, or measuring devices). By drawing and labeling this diagram, both teachers and students reinforce that students have processed and understood what they are to do. Similarly, students visually represent their learning following a lab (or other learning experience).

Teachers can use a similar process when students read the experimental design of another scientist, ranging from reviewing science fair procedures to reading from a professional science journal. This typically applies in a high school setting, possibly with an Advanced Placement (AP) science course. Ask students to read a journal article that includes experimental design and then construct a storyboard that outlines the procedure and materials the investigators used. This quickly shows that students understood what was presented, can describe the experiment, and can communicate key findings. This visual representation can include sketches, data representations, and brief segments of text.

Using Dramatic Instruction to Convey Critical Content

Emotions and unique experiences can play a key role in cementing science concepts in students' brains. Research has identified *interest* as an emotion that is key to learning, motivation, and social-emotional development (Silvia, 2008). You can generate interest in a variety of ways—by standing on a desk or table when making key points, or by using visual cues for a key concept you will relate, repeat, or come back to. Before doing a lab procedure that includes critical safety measures, for example, stand on a table, ensure all eyes are on you, and then demonstrate and describe what students need to know and understand. Students will be drawn in by the novelty.

Once the novelty captures attention, teachers can generate sustained interest by demonstrating what-if scenarios to highlight the dangers of not following safety protocols. For example, if you're talking about the importance of safety goggles, you can place a pair of goggles on a styrofoam head and splash some acid on it. Students will see how the acid deforms the styrofoam around the goggles but not where the eyes would be. Students will leave with an understanding of the importance of goggle wear that has emotional impact.

The options are limitless, so think creatively about the options that appeal to both you and your students. For example, when representing a historical scientist or even taking on a fictitious role related to a science concept, you can dress the part and stay in character. This could include having a classroom visit and question-and-answer session with Albert Einstein, George Washington Carver, Marie Curie, and so on. Students will go along with you as long as you are willing to stay true to the character. These activities provide memorable experiences and captivate students, particularly when they are not overused.

Element 17: Reviewing Content

Reviewing content provides students with the opportunity to recall material they previously learned and perhaps alter their thinking about it. As the science class moves along, it is important to ask students to pull forward concepts they learned earlier in the year (and in prior years). As the year progresses, learning tasks should continue to draw on and synthesize learning from earlier in the year.

Teachers must view learning as a continuous process if they are to develop lifelong learners. Students must understand that learning does not involve consuming discrete packets of information, taking a test, and then forgetting the information forever. If a teacher builds a curriculum thoughtfully, skills and concepts repeat throughout the year in new applications and scenarios. As old skills and concepts repeat, new information is incorporated to build ever-more expansive understandings.

Figure 6.3 shows the self-rating scale for reviewing content.

Score	Description
4: Innovating	I engage in all behaviors at the Applying level. In addition, I identify those students who do not have a correct and complete understanding of the previously learned content and design alternate activities and strategies to meet their specific needs.
3: Applying	I engage in activities to review content without significant errors or omissions and monitor the extent to which students have a correct and complete understanding of the previously learned content.
2: Developing	I engage in activities to review content with students without significant errors or omissions.
1: Beginning	I engage in activities to review content with students but do so with errors or omissions, such as not reviewing content that is important to the upcoming lessons and not making connections to broader concepts and generalizations.
0: Not Using	I do not engage in activities to review content with students.

Figure 6.3: Self-rating scale for element 17—Reviewing content.

We highlight the following strategies aligned to this element within this section.

- Cumulative review
- Demonstration
- Give one, get one

Cumulative Review

One of the strengths of the NGSS is that the science and engineering practices (SEPs) and crosscutting concepts (CCCs) run through *all* science experiences no matter what specific disciplinary core idea (DCI) students might learn (NGSS Lead States, 2013). At all times and in all science courses, teachers should continually refer to prior learning for SEPs, DCIs, and CCCs as part of cumulative review. If not referring to prior learning, students cannot build webs of understanding that can grow in complexity, increase in connections, and be modified. Students often feel this is redundant ("We learned that last year"); however, it is important for teachers to reinforce the curriculum concepts and build on the complexity and rigor students are asked to embrace.

For example, the concept of density is abstract and necessarily draws on prior learning. When learning about density, students should review concepts such as *mass is always conserved*. In fifth grade, students usually learn about properties of matter, including those characteristic physical properties that identify a particular substance. However, based on the conceptual challenge of density, it is usually reserved for middle school. Rather, fifth-grade students explore ideas such as color, hardness, conductivity, magnetism, and so on. Later, when students study the concept of density in middle school, they are reminded of the concepts they learned previously, including the idea that matter can't be created or destroyed, along with the other physical properties they learned. Figure 6.4 (page 84) shows a pair of performance expectations for a learning progression.

5-PS1-3.	Make observations and measurements to identify materials based on their properties.
MS-PS1-2.	Analyze and interpret data on the properties of substances before and after the substances interact to determine if a chemical reaction has occurred. [Clarification statement: Examples of reactions could include burning sugar or steel wool, fat reacting with sodium hydroxide, and mixing zinc with hydrogen chloride.] [Assessment boundary: Assessment is limited to analysis of the following properties: melting point, boiling point, solubility, flammability, and odor.]

Source for standards: NGSS Lead States, 2013.

Figure 6.4: Performance expectations for a learning progression.

One of the earliest topics in a physics course is kinematics. As students study motion, they are asked to interact with various metric measurements, such as the meter and kilometer, and relate their proportionality. Students must use proportional reasoning. This can include predictions and applications of these units to find how many meters it is to walk to their next class, drive home, fly to another city, and so on. As students do this, they also interact with relative distances and the unit that best represents these measurements. This reinforces and expands on earlier learning with these units of measurement in previous years and asks them to apply and combine units in new and novel ways.

Teachers should continually review and connect skills to prior learning and applications because much of science learning is thematic—specific concepts are common threads that run throughout the curriculum. For example, the concept of precision in measurement runs throughout science. Students are taught to measure using a variety of tools in their earliest science experiences. As students move along, teachers introduce the concepts of precision and accuracy in measurement, identify the difference between these two terms, teach students how to recognize the presence or absence of each, and provide practice to improve the precision or accuracy of measurement. In addition, teachers typically introduce students to the idea of significant figures in high school science, taking this to a new level in reporting measurements. Chemistry students are typically taught the basics of how to identify and report the correct number of significant figures. Teachers should review and apply this concept throughout the chemistry course and continue it in subsequent coursework, most often physics, where students can review and build on these concepts.

Demonstration

One of the most exciting things about learning and teaching science is the ability to observe concepts in action. Demonstrations are great tools for problem-based learning. Problem-based learning focuses students on three things: (1) understanding concepts, (2) understanding the principles that link concepts, and (3) linking concepts and principles to conditions and procedures for application (Gijbels, Dochy, Van den Bossche, & Segers, 2005). Demonstrations typically involve the evaluation of visually observable phenomena and can be conducted to meet some or all these conditions. Demonstrations can be teacher directed, teacher moderated, or completely student driven. There are many ways students can demonstrate their understanding, allowing them to review concepts and deepen their understanding. These experiences, in which students show that "I can do this," develop confidence in both knowledge and skill. For example, in the sludge test, a classic physical science experience, students demonstrate their understanding of physical properties and how they can use them to separate and identify substances. Students use the various characteristic physical properties, such as density, boiling point, magnetism, or flammability, to separate a mixture of solids and liquids, as well as identify individual samples once separated.

Similarly, students demonstrate their understanding of projectiles, forces, and kinematics by completing calculations to predict the flight of a projectile, and then test their predictions by actually launching the projectile. Teachers are able to generate excitement and confidence in students as they prepare for this experience, with the launch being an exciting and valid test of student understanding. These projectiles can be launched with a student-designed and built catapult or using one of the many commercially available launchers.

Give One, Get One

In this strategy, as the name implies, students both give ideas to and receive ideas from their peers. This strategy gets students talking with one another so they can serve as learning resources for each other. Research shows that group work cultivates both help-giving and help-receiving behaviors. Important help-giving behaviors include providing detailed explanations and sharing thought processes rather than bits of information (Webb & Mastergeorge, 2003). Examples of help-receiving behaviors include asking for specific explanations instead of answers, seeking explanations, and immediately applying peer-received help. Sharing via a give one, get one strategy is particularly helpful when exploring topics that have extensive vocabulary or concepts such as with the phases of the moon or cellular organelles. For example, at the end of a seventh-grade unit on the moon and its phases, students can write everything they know about the phases of the moon. This act of retrieval is important to activate thinking and prepare for interactions.

Once the list is in place, interactions can be structured in various ways. One approach is to use the Stand Up, Hand Up, Pair Up strategy (Kagan & Kagan, 2015), in which students stand with a hand up, identify another student looking for a partner, give a high-five, and then share one idea with each other. Each student takes one idea or concept from his or her partner and then rotates to another partner until he or she has compiled an extensive list. Students can then return to their seats, review the list of give ideas (what they shared) and get ideas (what they learned from others) and add them to their notebooks. This is a great way to review material, promote engagement, and cement understandings.

Element 18: Revising Knowledge

Revising knowledge is the overt act of creating opportunities for students to make changes or add to what they thought they understood. To allow students opportunities to revise knowledge, teachers must employ a system to maintain evidence of past student thinking. In some cases, teachers may ask students to maintain a portfolio of past work on paper. Periodically, you could ask students to take out prior work and make revisions or transfer past knowledge to a new context. You also may employ this strategy digitally through maintenance of digital folders that save a history of past work. Revisions can be a long-term process as just described, or revisions can happen within the length of single class period or lesson. Regardless, the process requires a reflective student moment, acknowledgment of past mistakes, time to make changes, and a guided way to bring past learning into the future.

Figure 6.5 (page 86) depicts the self-rating scale for this element.

The following strategies provide teachers with ways to help student facilitate the intentional revision of their knowledge.

- Peer feedback
- Assignment revision
- Writing tools

Score	Description
4: Innovating	I engage in all behaviors at the Applying level. In addition, I identify those students who are not making changes to their knowledge that enhance their understanding and design alternate activities and strategies to meet their specific needs.
3: Applying	Along with adequate guidance and support, I engage students in activities that help them revise their knowledge and monitor the extent to which students are making changes to their knowledge base that enhance their understanding of context.
2: Developing	I engage students in activities that help them revise their knowledge and provide adequate guidance and support.
1: Beginning	I engage students in activities that help them revise their knowledge but do not provide adequate guidance and support, such as reminding them to look for and correct mistakes, identify and fill in gaps in understanding, and examine the reasons behind the changes they are making.
0: Not Using	I do not engage students in activities that help them revise their knowledge.

Figure 6.5: Self-rating scale for element 18—Revising knowledge.

Peer Feedback

Providing and supporting experiences for students to give and receive peer feedback is a powerful learning strategy from which all students benefit. When giving feedback, students are able to compare peer work to their own and ascertain different levels of quality for the same assignment. When receiving feedback, students realize that no product is perfect, and different perspectives can be helpful to make that product better. With peer feedback, the stakes are low, since no grade is typically attached. Students understand that the feedback process provides a second chance to improve their work before assessment by the teacher.

This process requires all students involved to understand and reflect on the criteria established for judging successful work. Students must then compare these success criteria to the work sample in order to provide feedback to peers. For example, if students create a visual model to represent the cycling of nutrients in an ecosystem, their peers can review the model to identify both its strengths and limitations based on what they've learned to date. This can include looking at the details and accuracy of the model, how well the parts are labeled, and how the model can explain the transition in nutrients between ecosystem components.

One way for students to provide strong feedback on project-based tasks such as this is to have them incorporate technology tools. For example, students could record verbal feedback via phone, tablet, or other device. After recording, students can share audio files with peers.

This allows peers to provide feedback quickly and also allows the student receiving the feedback to easily retrieve it following class. Because asking students to formulate and coherently verbalize feedback is a skill we want students to develop, this process also develops skills for students providing the feedback.

Assignment Revision

Similar to peer feedback, it's important that students have opportunities to review and make changes to their first attempts at work. The process of doing science is not a one-and-done endeavor. Just like scientists continually refine their hypotheses and engineers iterate on design solutions, teachers must model for students the process of refining and revising their first attempts on an assignment. While there is certainly a balance in ensuring students put their best effort into their initial attempts, our experience is that students do not produce intentionally shoddy or incomplete work just so they can revise it later. Students understand that a solid initial attempt leads to less revision work in the long run. Ongoing refinements develop persistence, resilience, and a growth mindset (Dweck, 2016).

When providing feedback that drives assignment revision, it is important the feedback draws student thinking forward. This does not include platitudes such as "Good work," "Well done," or even "You could have explained more." Rather, the feedback should stimulate specific avenues for thinking and call students to concrete action. This could include statements or questions such as the following.

- "What is one additional piece of evidence you could include to support your claim?"
- "How could you clarify the connection between temperature and pressure?"
- "Tell me more about how the predator actually directly benefits the prey species in this environment."

Asking these types of questions about student work, including offering an opportunity to verbalize or write a response, pushes students to review their own work, consider the feedback, and then take action to revise the work in a way that shows growth. With the increasing use of digital tools, this becomes even more viable and easy for students to do.

Writing Tools

One of the most important skills students can develop in science is the ability to process and create informational text. Nonfiction writing is key to science understanding and appears in nearly all science and engineering practices, but it is directly called for in SEPs 3, 6, 7, and 8 (NGSS Lead States, 2013). To build these skills, it's important to provide scaffolds and prompts that give students practice in writing. We recommend using concept generalizations and stems for this purpose.

Concept Generalizations

In this approach, teachers provide students with several related terms, and students must generate a paragraph using all the terms. The paragraph students produce must demonstrate the relationship between the terms and not just the definitions. On the surface, this seems to be a straightforward and simple task, but it requires a depth of understanding about the relationships of these terms. For example, middle school standard MS-LS1-6 tells students: "Construct a scientific explanation based on evidence for the role of photosynthesis in the cycling of matter and flow of energy into and out of organisms" (NGSS Lead States, 2013). Teachers will certainly break down this performance expectation into various labs and activities. At various points in the lesson, they could ask students to generate a paragraph using the following terms to demonstrate their learning and current understanding: *photosynthesis, plants, carbon dioxide, oxygen, water,* and *glucose*. Even better, students can expand the explanation later to include the results of labs and activities done in class, such as showing the actual numbers for CO_2 generated from sources or the uptake of CO_2 by plants in the environment.

Stems

Another writing tool includes using *stems*, which are prompts for structures that students use to compose their thinking. Rather than simply asking an open-ended question, a stem provides a framework in which students can give a response. For example, with constructing scientific explanations or engaging in arguments from evidence, teachers usually expect students to follow a particular structure in their writing. This often includes a claim, evidence, and reasoning. It's helpful to provide a tool (such as the one in figure 6.6, page 88) to help students organize their thinking, particularly early on in the school year or during skill development.

Constructing Scientific Explanations or Engaging in Argument From Evidence	
Claim: Answers the question in an accurate, concise, and clear way.	
Evidence: Provides specific data from the reading to support the claim; you can include numbers or statements.	
Reasoning: Describes the meaning of the evidence or how it supports the claim, connects concepts and terms learned in class, and includes relevant vocabulary in the reasoning.	

Figure 6.6: Claim, evidence, and reasoning chart.

*Visit **go.SolutionTree.com/instruction** for a free reproducible version of this figure.*

Teachers can use this tool to highlight the success criteria students are striving for, help students organize their thinking, or establish a framework for providing students with feedback.

Element 19: Reflecting on Learning

Reflecting on learning involves students analyzing the extent to which they behave like effective learners. Good student reflections require both time and space for students to explore what they've learned well and uncover learning gaps. By creating safe spaces for students to engage in metacognition, teachers can develop self-directed learners who revisit past thinking and make revisions. Systematic reflection allows students to learn from both their successes and failures (Ellis, Carette, Anseel, & Lievens, 2014).

Figure 6.7 depicts the self-rating scale for element 19.

The following strategies are associated with this element.

- Exit slips
- Knowledge comparisons

Score	Description
4: Innovating	I engage in all behaviors at the Applying level. In addition, I identify those students who are not aware of their major impediments to learning and design alternate activities and strategies to meet their specific needs.
3: Applying	Along with adequate guidance and support, I engage students in activities that help them reflect on their own learning and monitor the extent to which students are aware of their major impediments to learning.
2: Developing	I engage students in activities that help them reflect on their own learning and provide adequate guidance and support.
1: Beginning	I engage students in activities that help them reflect on their own learning but do not provide adequate guidance and support, such as reminding them to continually monitor their level of understanding as well as their levels of effort and attention.
0: Not Using	I do not engage students in activities that help them reflect on their own learning.

Figure 6.7: Self-rating scale for element 19—Reflecting on learning.

Exit Slips

As noted in chapter 1 (page 11), the end of a learning segment should allow for individual reflection or a check for understanding. Exit slips are brief written (or digital) responses from students that demonstrate or reflect on their learning progress. They could be a question that summarizes learning from the day, a check for understanding, or a reflective moment. Exit slips are common in most classrooms. However, in some cases we have seen them used as a perfunctory end to class and not in ways to best increase student learning.

Teachers should be strategic with the use of exit slips. Science teachers might consider what information will be most helpful to them in planning future instruction. For example, consider how to use exit slips effectively if students in chemistry are writing chemical formulas, balancing chemical equations, and predicting products. Teachers could ask students a question or two that, in five minutes, provides insights on where they are succeeding or struggling. Some students might be stuck on the idea of writing an ionic formula. Others may have completely mastered formulas but are still grappling with the concept of predicting products. Still others may demonstrate some misunderstandings or a lack of confidence on the use of coefficients versus subscripts when balancing the equation. Based on student responses, teachers can develop working groups (or a lesson for all) for the following day, allowing intentional intervention and conversations that benefit students before proceeding to the next lesson.

Exit slips can also come in the form of a reflective prompt. This could include students' reflections on how they are progressing on a scaled learning target, such as their proficiency in constructing a scientific explanation or creating data representations such as graphs or tables. As students reflect on their mastery of the success criteria, they better focus their review or study for the coming time period. This can also serve as the basis for powerful conversations between a teacher and student about their respective assessments of proficiency and their plans for closing learning gaps.

Exit slips encourage students to pause, reflect, and retrieve. This is an important part of the learning cycle. As Lang (2021) writes in his book *Small Teaching*, there is great value in having students retrieve information from memory, which in turn solidifies that knowledge or skill for the student. Having students write down their thinking provides a testing effect, thereby better cementing this learning in place.

Knowledge Comparisons

When utilizing the knowledge comparison strategy, teachers create opportunities for students to connect old knowledge to new knowledge in a way that honors student experiences and exposes misconceptions (Rau, 2018). As science teachers, we tend to think of ongoing formative assessment most closely related to developing science skills. It's also helpful to measure the acquisition of science knowledge or deepening conceptual understandings.

For example, the *force concept inventory* (FCI) is a well-known, well-researched, and validated tool in physics classrooms. While it was developed in 1985, teachers continue to use it in physics classrooms to identify misconceptions held by students about force and motion (Bani-Salameh, 2016). This test provides insights about students' understanding of some of the basics of Newtonian physics in everyday language. The items on the FCI represent concepts typically taught in a first-semester physics course, specifically related to forces. This multiple-choice assessment is intentionally designed as a pre- or postassessment students can take in approximately thirty minutes. In particular, the FCI exposes many misconceptions that students bring with them to a physics class from their lived experience. Understanding these misconceptions is critical, and teachers need to know the extent to which these misconceptions have been overcome during the course of learning.

Element 20: Assigning Purposeful Homework

Teachers frequently use homework, but not always in ways that enhance student learning. Assigning purposeful homework requires that the teacher plans the unit thoughtfully in advance with the final learning outcomes in mind. In the sequence of learning, teachers should only assign homework if it has a specific rationale that supports the learning in a way that could not be handled during the standard class time. With homework, in general, less is more. In addition, homework should not inordinately affect individual student grades because student ability to complete the work may depend on a variety of factors (for example, student employment, lack of parent support at home, access to technology) that do not relate to student proficiency. Since homework presents both equity and efficiency challenges, homework must be purposeful and accessible for all students.

Figure 6.8 shows the self-rating scale for this element.

4: Innovating	I engage in all behaviors at the Applying level. In addition, I identify those students for whom the assigned homework does not enhance learning and design alternate activities and strategies to meet their specific needs.
3: Applying	I engage in activities to assign purposeful homework without significant errors or omissions and monitor the extent to which the homework enhances student learning.
2: Developing	I engage in activities to assign purposeful homework without significant errors or omissions.
1: Beginning	I engage in activities to assign purposeful homework but do so with errors or omissions, such as assigning homework not directly related to the critical content addressed in class and assigning homework for which students are not adequately prepared.
0: Not Using	I do not engage in activities to assign purposeful homework.

Figure 6.8: Self-rating scale for element 20—Assigning purposeful homework.

In this element, we focus on the strategy of homework to practice a process or skill.

Homework to Practice a Process or Skill

An ongoing challenge for all teachers is to consider how they ask students to engage in learning that continues beyond the classroom time and space. It's important that this outside work is both purposeful and equitable. Certainly, students have different levels of support and materials once they leave class, as well as different levels of motivation and executive functioning skills to complete the tasks. Although teachers may believe a student has capacity for independent work based on what they see in class, the home environment may not be conducive to achieve the same results.

For example, students with ADHD and learning disabilities need focused support at home. Parents are not typically universally equipped with the same skill sets to support their children's learning at home. Providing guidance can help parents partner with teachers to support learning and get more involved with their children's learning. Allison R. Walker, Terri S. Collins, and Amelia K. Moody (2014) provide suggestions on structuring home learning environments, teaching children self-monitoring strategies, and providing supports for learning at home.

In this book and our daily work, we continue to focus on moving thinking about science learning from "a bunch of stuff people should know" to viewing science as a process of doing and applying science skills. Similarly, it is helpful to provide homework tasks that are engaging, practice science skills from class, and have multiple answers, thereby cutting down on the tendency or temptation to simply copy from a source or another student.

For example, if students are learning about measurement or the metric system, have them think of five items they will measure when they get home. This will, most likely, involve length or volume, with students potentially taking a measuring tool from the classroom. Students estimate what they believe the measurement will be and then come back the following day with the actual measurements. This task could be expanded to include measuring items related to a hobby students love or a tradition in their family, motivating them to bring other parts of their identity to the classroom. In addition, given the proliferation of device-embedded cameras in phones and tablets, students can also take a picture of the item and the measurement they recorded. This could lead to conversations such as how the estimate compared to the actual measurement or sharing about the items themselves.

Another task might relate to learning about classification systems, like in a life science class. Just like the biological system of classification and binomial nomenclature, there are many classification systems for other items. Students could develop a hierarchical system for classifying household items, ranging from socks to kitchen utensils to tools. Again, this process allows students to bring aspects of their identity in the items they choose for homework and then share with the class. Students could focus on the skill of classification as well as construct an explanation of their organizational system and the characteristics used to differentiate items into particular categories. A further extension of this could include viewing this system as a model (SEP 2) and addressing the strengths and limitations of the model (National Research Council, 2012).

Element 21: Elaborating on Information

Elaboration is an inferential act that results in the creation of new awareness when students engage in it effectively. When students elaborate, they go beyond what they initially learned to explore new aspects of understanding. There are myriad possibilities when elaborating on skill development in the sciences. Elaboration may involve exploring a concept in more depth, increasing the number of conceptual connections, or expanding current understandings to new contexts.

This section presents strategies for elaborating on information in science classes. Figure 6.9 (page 92) depicts the self-rating scale for this element.

Score	Description
4: Innovating	I engage in all behaviors at the Applying level. In addition, I identify those students who are not making substantive additions to their knowledge base and design alternate activities and strategies to meet their specific needs.
3: Applying	Along with adequate guidance and support, I engage students in activities that help them elaborate on information and monitors the extent to which students are making substantive additions to their knowledge base.
2: Developing	I engage students in activities that help them elaborate on information and provide adequate guidance and support.
1: Beginning	I engage students in activities that help them elaborate on information but do not provide adequate guidance and support, such as not sequencing questions in such a way as to gradually increase the rigor of students' responses and not pushing students to expand on their answers.
0: Not Using	I do not engage students in activities that help them elaborate on information.

Figure 6.9: Self-rating scale for element 21—Elaborating on information.

For this element, we focus on the strategy of elaborative interrogation.

Elaborative Interrogation

At times, teachers (and students) can become much more focused on the procedures and processes associated with carrying out the task and miss the learning point altogether. In these situations, questions and instructions can become more procedural in nature, simply prompting students on how to understand and comply with lab instructions. This makes laboratory investigations in the science classroom involving interrogation one of the best ways to have students interrogate ideas and elaborate on what they know. Through such elaborative interrogations, teachers continually refocus their thinking and students' thinking on the things they want students to *learn* as a result of doing the lab.

A good elaborative interrogation involves thoughtful questions that require students to apply their prior knowledge (personal experience, pre-lab procedure, and background reading) and incorporate ideas that are emerging in the laboratory investigation (from observation and data collection). This moves thinking from a focus on *how* to do the lab to *why* to do the lab.

To do this well, teachers should plan questions in advance, write them down, and use them while moving between lab groups. Writing questions in advance allows teachers to refine, remember, and share these questions with colleagues or improve them throughout the day. While walking between lab groups, teachers can check in and ask students one of the following questions.

- "What happened?"
- "What evidence do you have that it happened?"
- "Why did that happen (science reasoning)?"
- "What might have happened if you _____ instead of _____ ?"

These and similar questions, when used in a timely and strategic manner, push students to think differently and more deeply as they process lab observations. These questions may simply be posted by the teacher while he or she circulates around the room. Skillful teachers will time their questions, pose them to a student group, listen for a conversation to begin, and then move on to the next group. It is important, however, for teachers to cycle back and recheck students' understanding, both to maintain integrity in the questioning process as well as ensure understanding. In addition, the teacher may stop the laboratory investigation temporarily, pose a question, and then collate student answers on the board. Following this, the teacher may lead a discussion exploring which responses best answered the question and solicit suggestions for revisions.

Ultimately, we encourage teachers also to have students generate questions that fit this pattern and push them to think differently. This approach is especially well-suited to phenomena-based investigations. When presented with a novel phenomenon, students can ask questions that are testable and then apply them to designing an investigation to find a solution. This brings student ownership and responsibility to the learning process, as well as supporting growth in one of the science and engineering practices: asking questions and defining problems.

Element 22: Organizing Students to Interact

Teachers should organize students into collaborative groups thoughtfully and deliberately to ensure that they interact effectively to enhance their learning. This is an important part of science learning, whether promoting collaborative conversations or working efficiently in the laboratory setting. In "Cooperative Learning: Principles and Practice," Jason Anderson (2019) indicates that smaller groupings (three to four students) maximize engagement. In addition, a great deal of care should be put into how these groups are structured, citing the benefit of both homogeneous and heterogeneous groups. The key idea is to be thoughtful in how the groups are formulated and then engaged to support the learning of each student (Anderson, 2019). Based on context and situation, a teacher may stray from these suggestions and experiment with a variety of group sizes and ability distributions.

This section presents five strategies for organizing students to interact in the science classroom. Figure 6.10 depicts the self-rating scale for this element.

Score	Description
4: Innovating	I engage in all behaviors at the Applying level. In addition, I identify those students who are not actively involved and interacting deeply with the content and design alternate activities and strategies to meet their specific needs.
3: Applying	I engage in activities that organize students to interact without significant errors or omissions and monitor the extent to which students are actively involved and interacting deeply with the content.
2: Developing	I engage in activities that organize students to interact without significant errors or omissions.
1: Beginning	I engage in activities that organize students to interact but do so with errors or omissions, such as failing to establish protocols for interaction and lacking a clear purpose and goals for interacting.
0: Not Using	I do not engage in activities that organize students to interact.

Figure 6.10: Self-rating scale for element 22—Organizing students to interact.

For this final element pertaining to all types of lessons, we focus on the following strategies related to science instruction.

- Group norms creation
- Group using preassessment information
- Think-pair-share and think-pair-square
- Peer-response groups
- Group reflection on learning

Group Norms Creation

Students in the science classroom work together frequently, and successful peer interactions are critical for students' science learning. To ensure effective interactions, teachers and students should work together to define what successful interactions look like. These norms serve as the collective commitments for each member of the classroom community. We recommend generating a list of observable behaviors by first prompting students' thinking with questions such as: "Think back to your most successful and enjoyable classroom experience. Consider both the learning that took place and the enjoyment you experienced. What did you do to contribute to that positive experience? Be as specific as you can. What did the teacher do to contribute to that positive experience? Be as specific as you can."

Allow students to think individually, compare and refine thoughts in small groups, and then bring their ideas to a class discussion. Once the teacher and students compile, refine, and agree to the list of group norms, the teacher may display them in the classroom by, for example, printing them on 11" x 17" paper, asking all students to sign the paper, and then displaying it for ongoing reference, reminders, and celebration. You can use this process, with some slight modifications, for lab groups that work together over an extended time period. For example, a biology teacher may assign a specific lab grouping that works together a three-week unit on ecology. During this time period, the group may need to construct written norms and identify roles to work effectively with the sequence of labs conducted throughout the unit.

Group Using Preassessment Information

Planning lessons and groupings using students' prior experiences, expertise, comfort levels, and knowledge both honors where students are and leads to a more efficient learning experience. By taking all this information into account, teachers can place students in groups with defined purposes and highlight individual student strengths through assigning roles. We find that preassessments are often best given a day or two before forming groups. In addition to giving the teacher time to form the groups, this process also serves to model the teacher's positive response to student voice and background. Teachers can use the results of these preassessments to form homogeneous or heterogeneous groups. Most important, the teacher should share the rationale for group formation with asset-based language that honors the interests, skills, and talents of individual students.

With *heterogeneous groupings*, students who are more proficient or more confident can support and help those who are still struggling with a concept or a skill. With *homogeneous groupings*, the teacher can group learners of similar proficiency together, taking specific care to group emerging learners together to intentionally work with them to bring them up to speed or to increase their confidence. Teachers then place more confident learners into an enrichment experience that pushes them to think in new and different contexts. As such, the practice or task the teacher gives to various groups is differentiated or scaled in response to the preassessment data.

Grouping students based on preassessment does not need to be limited to knowledge assessments. In addition to seeing what students know about a particular science topic, preassessment groups could be based on confidence rankings. For example, when working with microscopes, Bunsen burners, or even constructing a lab procedure, some students feel extremely confident with the necessary skills, while others may share a lack of confidence or experience. Again, teachers can structure scaffolding so students receive different tasks or different structures based on what they share with the preassessment instrument.

Think-Pair-Share and Think-Pair-Square

One of the most widely used and effective classroom strategies is think-pair-share. Specific research on think-pair-share in chemistry classrooms showed significantly higher test scores than a traditional lecture strategy (Bamiro, 2015). In a think-pair-share, the teacher presents a problem or task to the class. Students have to individually process the question and consider how they might respond. Students then turn to a predetermined partner, often a shoulder partner, to pair and discuss their responses. This process can serve to refine and expand each student's thinking in a safe and engaging context. Finally, students share their thinking with the class. Having already checked and verified answers with a peer, more reserved students are often more motivated to share with the larger group. This can be combined with other strategies, such as response boards (as discussed in element 4, page 23). For example, a teacher could ask students to perform a chemistry calculation, complete a ranking task in physics, or present an argument for a position such as, "Should all humans eat a vegetarian diet to curb climate change?" No matter the format, these conversations will engage students more than putting the prompt before the class and then calling on volunteers.

An extension of think-pair-share is think-pair-square. In this approach, instead of a representative from a pair sharing with the entire class, two pairs form a group of four to share their thinking with one another. For example, two student lab groups could conduct an investigation focused on the same independent variable. They could then combine and compare their results, essentially doubling the number of trials they are able to analyze. Alternately, lab groups could investigate different independent variables, compare their data, determine which variable was most impactful, and so on, leading to a more jigsaw-like learning experience. Again, as with think-pair-share, the goal is to get small groups of students actively thinking, reflecting, and talking about the science concepts they are studying.

Peer-Response Groups

When students work with peers to give and receive feedback on complex tasks, not only are students engaging in conversations, but these conversations serve to deepen students' understanding of science concepts and skills. Deep learning results from acts of reciprocity, and knowledge can be co-constructed through discussion of the validity of claims (Bamber & Crowther, 2012). Peer response groups facilitate both of these. In peer response groups, students come to better understand the success criteria. In element 1 (page 11), we highlighted the value of using teacher-generated scales and success criteria. When these are established for recursive skills, such as outlined by the science and engineering practices, students can reflect on and evaluate their own as well as peer work samples.

For example, students can engage in conversations with peers using the scaled target, success criteria, and co-constructed feedback tool in figure 6.11 (page 96). Teachers provide structures and roles as appropriate. In this example, the teacher asks students to use five pieces of their work spanning the spectrum of proficiency to see varying levels of quality between work samples. Students reflect on the questions in the center column of figure 6.11 related to their own work, find evidence of where the work does or does not meet the success criteria, and then provide a score to each work sample. Once each student in the group provides a score for his or her own work, group members can share, discuss, and defend the assigned score using the Comments and Feedback column.

Teachers who want to shift the focus from self-assessment to peer assessment can modify the center column, Student Reflections, to focus on providing feedback to a peer.

4	3	2	1
I can construct an explanation for a scientific phenomenon using all success criteria in unfamiliar contexts or making connections to related science concepts.	I can construct an explanation for a scientific phenomenon using all success criteria in familiar contexts.	I can construct an explanation for a scientific phenomenon using some success criteria in familiar contexts.	I can construct an explanation for a scientific phenomenon in familiar contexts with support.

Success Criteria:

- Create an accurate claim.
- Use multiple valid and reliable sources of evidence.
- Provide clear and complete reasoning that supports the claim clearly connects with science concepts and vocabulary (these can be listed or not; for example, *plant, oxygen, carbon dioxide, sugars, light energy, flow, leaf, photosynthesis, cycle*).
- Clearly demonstrate a connection to the crosscutting concept

Student Reflections and Co-Constructed Feedback		
Learning target: I can construct an explanation for a scientific phenomenon using all success criteria in familiar contexts.		
Success Criteria for Constructing a Scientific Explanation	**Student Reflections: Examples From Your Work**	**Teacher (or Peer) Comments and Feedback**
Create an accurate claim.	Review your claim. Did it make an accurate qualitative or quantitative claim?	
Use multiple valid and reliable sources of evidence.	What sources of evidence did you use to support your claim? How do you know these data sources are reliable? Is there any additional evidence you could include?	

Provide clear and complete reasoning that supports the claim and clearly connects with science concepts and vocabulary (these can be listed or not).	Is there a clear link between the evidence and what you learned in class? What references did you or could you make to the textbook, class notes, or other activities? Did your reasoning clearly explain the phenomenon or consider unanticipated effects?	
Clearly demonstrates a connection to the crosscutting concept.	Did you clearly identify the crosscutting concept in your explanation?	

Figure 6.11: Scaled target for constructing scientific explanations.

*Visit **go.SolutionTree.com/instruction** for a free blank reproducible version of this figure.*

Group Reflection on Learning

The world of science is rich and diverse. The more opportunities students have to reflect and form connections when learning science, the more likely they will gain a larger picture of science practices and how crosscutting concepts connect and weave in and out of the concepts they are learning. Group reflection allows students to highlight these connections. For example, students may recognize how the crosscutting concept of structure and function is evident in biology concepts such as photosynthesis, biochemistry and enzymes, and gene expression. While the connections are nearly endless, students' creativity and insights might bring to light a myriad of connections.

Lang (2021) describes the Minute Thesis as a strategy for engaging students in group reflection. The *minute thesis* is a thinking exercise in which the teacher presents a "big picture" idea along with associated smaller concepts. The teacher leads students through a series of activities in which the goal is to make conceptual connections supported by evidence. While Lang (2021) uses this strategy in an English language arts context, it also works for science. Teachers might do this at the end of a semester, but they could also do it at the end of a unit of study or other period. The numbers of connections are nearly limitless. As students make meaning and connections with this strategy, rich conversations and connections deepen students' learning. There are various possible approaches and modifications, but we will describe just one approach.

Figure 6.12 includes three columns, and each is independent of the other so it can be connected with others as part of the exercise. Students can make connections between the three columns using a variety of combinations.

Unit Example (Biology)	Science and Engineering Practice (SEP)	Crosscutting Concept (CCC)
Photosynthesis	Asking questions (and defining Problems)	Patterns
Biochemistry and enzymes	Developing and using models	Cause and effect
Cellular respiration	Planning and carrying out investigations	Scale, proportion, and quantity
Ecology and the impacts of humans	Analyzing and interpreting data	Systems and systems models
Homeostasis and regulation	Using mathematics and computational thinking	Energy and matter
Gene expression and inheritance	Engaging in argument from evidence	Structure and function
Evolution	Obtaining, evaluating, and communicating information	Stability and change

Figure 6.12: Making meaning and connections to reflect on learning.

Students would begin in small groups, each assigned to a specific unit (for example, those listed in figure 6.12). The groups connect that unit example to SEPs and CCCs while incorporating unit concepts. One student circles or identifies one item in each of the three columns. The other students then take one minute to determine what connections they can make between that unit of study (column one) with the SEP (column two) and the CCC (column three). Students can show their connections on butcher paper or whiteboards (huddleboards).

Following small-group work, the teacher may choose to lead a whole-class discussion while making explicit connections on the class whiteboard. In this manner, students build connections that later intersect with the work of other groups.

Guiding questions can stimulate thinking, such as the following.

- "How could you use _____ (SEP) to demonstrate the idea of _____ (CCC) in the _____ (unit)?"
- "How could _____ (CCC) be seen when you _____ (SEP) when learning about _____ (unit)?"

Individual students and groups can share their thinking with their group or class regarding how they would connect the three circled column items. Students take turns sharing or expanding on other students' ideas. As long as the idea holds true to the science concept (unit of study), the SEP, and the CCC, the teacher can commend students' thinking and then move on to the next idea. After sharing ideas, another student can take a turn identifying one item from each of the three columns and repeat the process. This activity can extend for any time period, and the review can make new connections and powerfully cement these concepts in students' minds.

GUIDING QUESTIONS FOR CURRICULUM DESIGN

For using strategies that appear in all types of lessons—direct instruction, practicing and deepening, and knowledge application—this is the design question: *Throughout all types of lessons, what strategies will I use to help students continually integrate new knowledge with old knowledge and revise their understanding accordingly?* Consider the following questions, which align to each of the eight elements in this chapter and guide teachers in planning instruction.

- **Element 15:** How will I help students preview content?

_____

_____

_____

_____

_____

_____

- **Element 16:** How will I highlight critical information?

_____

_____

_____

_____

_____

_____

- **Element 17:** How will I help students review content?

_____

_____

_____

_____

_____

_____

- **Element 18:** How will I help students revise knowledge?

_____

_____

_____

_____

_____

_____

- **Element 19:** How will I help students reflect on their learning?

_____

_____

_____

_____

_____

- **Element 20:** How will I use purposeful homework?

_____

_____

_____

_____

_____

- **Element 21:** How will I help students elaborate on information?

_____

_____

_____

_____

_____

- **Element 22:** How will I organize students to interact?

_____

_____

_____

_____

_____

Summary

Although theoretically, teachers could incorporate all of the aforementioned strategies in every kind of lesson, this would not necessarily be wise practice. Rather, teachers should use their professional judgment to judiciously balance the use of these strategies to help students continually integrate new knowledge with old. As students learn science throughout the year, it's helpful to bring novelty to the way students process their learning, but also to have consistency in strategies so students become comfortable and efficient in their groups. In the next chapter, we begin the category of *context* and discuss using engagement strategies.

PART III
Context

CHAPTER 7

Using Engagement Strategies

As stated in the introduction (page 1), *context*—the third of the three overarching categories (feedback, content, and context)—refers to students' mental readiness during the teaching–learning process. For students to be ready, teachers should work to meet their needs relative to engagement, order, sense of belonging, and personal interests and skills via learning profiles (Tomlinson, Brighton, Hertberg, Callahan, Moon, Brimijoin, Conover, & Reynolds, 2003).

Many educators use the term *engagement*; however, this term does not always have a clear definition. In fact, educators ascribe a wide variety of meanings to the term. For example, some educators might use the term to mean the simple behavior of paying attention to a lecture or demonstration in class. Others might use the term to refer to students being intrinsically motivated due to the content connections to their passions and interests (for example, exploration, hands-on laboratory work, scientific research, or connections to future employment possibilities).

The New Art and Science of Teaching (Marzano, 2017) addresses engagement from four perspectives. One is the traditional notion of *attention*. That is, some of the elements are designed to ensure students attend to what occurs in the classroom. A second perspective is *energy level*. Some elements involve strategies designed to increase students' energy levels, particularly when those levels are getting low. A third perspective is *intrigue*. Some of the elements address techniques to help stimulate high levels of student interest so that students seek further information about science content on their own. The fourth perspective is *motivation and inspiration*. Collectively, these perspectives include strategies that spark students' desire for self-agency and propel them to engage in tasks of their own design and control.

The following elements focus on the four perspectives of engagement.

- **Element 23:** Noticing and reacting when students are not engaged
- **Element 24:** Increasing response rates
- **Element 25:** Using physical movement
- **Element 26:** Maintaining a lively pace
- **Element 27:** Demonstrating intensity and enthusiasm
- **Element 28:** Presenting unusual information

CONTEXT

- **Element 29:** Using friendly controversy
- **Element 30:** Using academic games
- **Element 31:** Providing opportunities for students to talk about themselves
- **Element 32:** Motivating and inspiring students

Element 23: Noticing and Reacting When Students Are Not Engaged

Interest levels can be difficult to assess, as they change based on context and can shift from day to day. Within the course of a lesson or class period, students may be highly engaged in one sequence and off-task later on. The first step in addressing student engagement is to recognize when students are not engaged and react accordingly. When noting a lack of engagement, you may respond with a variety of strategies that shift from the status quo. These changes may involve things like including physical movement, asking unique questions, and changing pace or intensity.

Figure 7.1 depicts the self-rating scale for this element, which focuses on attending to students when they are not engaged.

Score	Description
4: Innovating	I engage in all behaviors at the Applying level. In addition, I identify those students who are not re-engaging and design alternate activities and strategies to meet their specific needs.
3: Applying	I engage in activities to notice and react when students are not engaged without significant errors or omissions and monitor the extent to which students re-engage based on their actions.
2: Developing	I engage in activities to notice and react when students are not engaged without significant errors or omissions.
1: Beginning	I engage in activities to notice and react when students are not engaged but do so with errors or omissions, such as focusing on only a few students when checking on student engagement and not reacting in a timely manner when students are disengaged.
0: Not Using	I do not engage in activities to notice and react when students are not engaged.

Figure 7.1: Self-rating scale for element 23—Noticing and reacting when students are not engaged.

For this element, we focus on the strategy of re-engaging individual students.

Re-engaging Individual Students

When teachers notice that students are disengaged, they should collect information and communicate with students to determine the cause of their disinterest. Taking the time to acknowledge when something is not working well for students shows respect and contributes to relationship building.

As an example, when students attempt to understand and communicate recursive, complex scientific content, their interests might wane, or the challenging nature of the work might lead to some cognitive roadblocks—so teachers need to reignite or provoke interest to re-energize and engage them. An easy way to check in with students involves simply asking them how things are working via Google form questions or using *mood meter* cards related to the task at hand (cards with emojis or other icons representing excited, neutral, bored, or similar emotions). In addition, you may be able to anticipate moments when engagement might falter and intentionally embed strategies to keep students motivated. For example, teachers may allow students to pick an experimental variable in a lab, design the experiment, and then jigsaw the conclusions based on the different experimental variables tested.

Brief brain breaks are also effective when working through long periods of difficult work (for example, stand up and walk around for one minute, watch a brief humorous video, engage in team-building games unrelated to science content). Brain breaks that involve physical movement have been correlated to positive academic and health outcomes (Martin & Murtagh, 2017). Teachers can employ any of these, as well as ask students what they might want to do to foster their own engagement. Inviting students to redirect or amend an initial plan or activity can help revitalize and invest them in the work ahead.

Element 24: Increasing Response Rates

In the science classroom, all student voices should be represented in the learning experience. This means that teachers need to create safe spaces for students to engage, especially when they are struggling with scientific content or skills. Therefore, teachers should create a variety of opportunities for students to engage that are less threatening than asking a question in front of the class (or responding orally to one posed by the teacher). In addition, if only a few students engage and the rest of the class remains silent, this may send the inadvertent message that the teacher only values the most outgoing or talkative students. Increasing the number of students who respond to a single question can greatly enhance the engagement level of the class as a whole (Heaslip, Donovan, & Cullen, 2013).

Figure 7.2 depicts the self-rating scale for this element.

Score	Description
4: Innovating	I engage in all behaviors at the Applying level. In addition, I identify those students who are not generating thoughtful and accurate responses and design alternate activities and strategies to meet their specific needs.
3: Applying	I engage in activities to increase response rates of students without significant errors or omissions and monitor the extent to which students are generating thoughtful and accurate responses.
2: Developing	I engage in activities to increase the response rates of students without significant errors or omissions.
1: Beginning	I engage in activities to increase the response rates of students but do so with errors or omissions, such as not providing adequate time for students to respond to the activity and not acknowledging that students have responded.
0: Not Using	I do not engage in activities to increase the response rates of students.

Figure 7.2: Self-rating scale for element 24—Increasing response rates.

The following strategies relate to increasing students' response rates.

- Hand signals
- Response cards
- Response chaining

Hand Signals

The use of hand signals is an extremely effective strategy to ensure all students are engaged in a line of questioning or skill development activity. Hand signals allow students to express a response in a physical manifestation instead of an oral answer. Oral answers in the whole-class setting can be embarrassing for some students, and hand signals provide a way to interact that is less intimidating. In addition, when asked to use hand signals, the entire class is typically doing it at the same time, and this involves *everyone* instead of individuals, who might be too shy to speak out. In addition, they allow the teacher to gather information regarding individual student understanding and make real-time instructional changes. Hand signals are simple, easy to use, and can be modified to achieve a variety of different purposes. At their simplest, students can use thumb signals in conjunction with a teacher prompt that asks them to discriminate in an either/or fashion. For example, a thumbs up may indicate *yes* or *true*. A thumbs down may indicate *no* or *false*. For multiple-choice or forced-choice prompts, teachers ask students to hold up a number of fingers to represent their answer (aligned with the number of choices presented—typically four to five). A fist indicates when a student doesn't know and requires assistance prior to answering (in a forced-choice scenario). Students can also use hand signals to indicate a perceived level of mastery. After teaching a complex concept, a teacher may ask students to hold up a certain number of fingers (for example, on a scale of one to ten) to indicate their confidence in their knowledge or skills.

For example, a high school chemistry teacher may want students to work individually on a series of multiple-choice questions like the following to assess facility with gas law concepts and calculations.

1. Which of these changes would cause an *increase* in the pressure of a gaseous system?
    a. Make the container larger.
    b. Add additional amounts of the same gas to the container.
    c. Decrease the temperature.
    d. Remove gas from the container.

2. A gas occupies a volume of 0.20 L at 76 mm Hg. What volume will the gas occupy at 760 mm Hg? (*The temperature stays constant.*)
    a. 38 L
    b. 20 L
    c. 2.0 L
    d. 0.020 L

3. A gas occupies 40.0 L at 123°C. What volume does it occupy at 27°C if the *pressure stays constant*?
    a. 182 L
    b. 30.3 L
    c. 80.0 L
    d. 20.0 L

However, prior to presenting these question to students, the teacher asks them to display their personal confidence in gas law concepts and calculations on a one to ten scale using hand signals. After gathering this preassessment information, the teacher displays the questions, one at a time, on an overhead or digital projector, allow thinking and calculation time, and then asks students to give hand signals (one to four fingers or fist) to respond. After each question, the teacher assigns mixed groups of correct, incorrect, and unsure students to promote collaboration to achieve the correct answer. If any of the questions led to all correct answers based on the hand signals, the teacher knows he or she can proceed to a subsequent learning activity or lesson without creating mixed groups. In other cases, after proceeding through the entire series of questions, the teacher repeats the confidence questions and assesses how students' responses have changed.

Note that when using this strategy, pre- and postassessment hand signals are critical for teachers to see if the associated learning activity accomplished learning goals. They also serve as visual reminders to students of how collaboration helps in the learning process as they see their growth as a class represented in the post-assessment signals.

Response Cards

Response cards offer a more flexible form of engagement than hand signals, as teachers can use them in conjunction with open-ended questions. Instead of using simple representations (finger, fist, open hand), response cards allow more complex communication involving a wider range of modalities. Response cards may be textual, graphical, or numeric in nature. Typically, students use small, individual whiteboards (for example, 12" × 12") and dry-erase markers. After an open-ended prompt (for example, "How does a tree in a pot become so much more massive than its seed if the mass of the dirt remains constant over time?"), students record their answers and display them to the teacher (or whole class) at the same time. Teachers may take note of the diversity of answers and plan instructional modification. Alternatively, teachers may group students and ask them to collaborate and revise their individual answers on a larger, shared group whiteboard (or huddleboard). In this manner, response cards (response boards) are an ideal way to get individual commitments from each student and then transfer those ideas and answers to collaborative thinking activities.

Response Chaining

Whole-class questioning techniques can be difficult to implement without prior planning. In whole-class questioning scenarios, many teachers will ask a student a question, receive a correct or incorrect answer, and then explain the error or provide praise. This type of teacher-centered questioning cycle engages only one student at a time. The approach also leads to some anxiety for students and possible shame for those who give incorrect answers. Response chaining puts the teacher in the role of mediator and facilitator. After the teacher asks one student a question and that student provides an answer, the teacher asks a second student to evaluate the response, noting items that are correct or incorrect.

The teacher may further redirect those responses to other students and ask related questions that help clarify a big-picture concept. Through redirection, students may come to realize that answers to some questions in the chain may be correct in certain contexts and incorrect in others. Continuous directed and redirected questioning requires students to listen to others, incorporate past answers, and modify their responses based on newly presented contexts. Once students can move in and out of different contexts fluidly, the big picture comes into focus.

A variation on this theme may involve the teacher asking students to verbally chain together ideas that relate to a larger process or scenario. For example, in a cellular unit, a middle school or high school biology teacher may ask the class how a B-cell produces an antibody and releases it into the bloodstream. Based on previously acquired knowledge of individual organelle function, students should be able to puzzle it together with guiding questions from the teacher, such as, "Where would this process start?" A student responds with "the nucleus," because that is where all protein-making instructions reside. The teacher then redirects and asks another student to propose where the next cellular location in the process would be. That student might respond with "the ribosome" because the student knows that mRNA copies of the DNA go to the ribosome to make the protein. If students make an error, the teacher redirects the question until the correct answer is uncovered. As a variation on this strategy, the teacher might direct this activity like the game of telephone, in which each student response must repeat all the prior steps (in this case—organelle location and function) until they explain the entire sequence of events.

Element 25: Using Physical Movement

In many science classes, students spend the duration of the lesson or class period seated as they work through a laboratory investigation or class activity. As we learn more about human biology, we have discovered that students need more physical movement to optimize learning. Physical movement introduces novelty and variety, and it stimulates blood flow to the brain (Marzano, Pickering, & Heflebower, 2011). By getting on their feet and navigating through the room, students also are better able to collaborate in different groups, interact with materials at different stations, view examples of peer work, and construct physical models to support learning.

These classroom journeys can provide students with different perspectives, challenge their assumptions, and let them approach future tasks with newly gathered insights. Generally, we've found that students return from periods of physical movement with renewed energy. With punctuated periods of renewed energy, we also have found that higher overall engagement levels can be sustained throughout longer stretches of classroom time.

Figure 7.3 depicts the self-rating scale for this element, which focuses on incorporating physical movement into the classroom.

Score	Description
4: Innovating	I engage in all behaviors at the Applying level. In addition, I identify those students who do not exhibit increased energy levels and design alternate activities and strategies to meet their specific needs as they relate to physical movement.
3: Applying	I engage in activities to increase the physical movement of students without significant errors or omissions and monitor the extent to which students exhibit increased energy levels.
2: Developing	I engage in activities to increase the physical movement of students without significant errors or omissions.
1: Beginning	I engage in activities to increase the physical movement of students but do so with errors or omissions, such as not using physical movement at times when students need an energy boost and not providing an appropriate amount of time for the physical activities.
0: Not Using	I do not engage in activities to increase the physical movement of students.

Figure 7.3: Self-rating scale for element 25—Using physical movement.

This section explores the following strategies related to using physical movement.

- Body representations
- Drama-related activities

Body Representations

To visually assess the variety of student opinions regarding a complex question, it is often helpful to ask students to *vote with their feet*. A simple version of this activity may involve posting answers on separate parts of a classroom wall or whiteboard and asking students to stand in front of the one they believe is most correct or entirely true. Although this gets students up and moving, this particular version of the activity does not offer the teacher much more than what the previously discussed hand signals could provide.

A more interesting version of this activity involves the teacher setting up the room as two polar opposites in relation to a particular question. After the teacher poses the prompt or question, he or she asks students to move in relationship to the ends of the spectrum to demonstrate the location of the correct answer. For example, a middle school physics teacher may pose the question, *How much does increased mass affect the acceleration with which a bowling ball falls from a building?* The teacher may tell students that one end of the room represents *quite a bit*, the middle of the room represents *some*, and the other end of the room represents *not at all*. The teacher then sets a timer for a brief period (maybe ten seconds) and says, "Go!" Students then move to the area they believe best represents the answer. When students stop, they will likely be in different places or clustered in different groups. Now that students have voted with their feet, the teacher can follow up with whole-class questioning to discuss their reasoning and compare prior beliefs, misconceptions, and new connections. If appropriate, the teacher may ask the same question after the discussion (if students have not identified the correct answer) and have students move again to see if they change their answers.

Drama-Related Activities

To assist in elucidating complex relationships like cause and effect, teachers may lead students in dramatizing (or acting out) phenomena or serving as actors in metaphor scenarios. Full-fledged acting can be an intimidating activity, especially for shy students. However, using student volunteers in a demonstration can be fun for those involved and engaging for the student audience. An example in a ninth-grade biology class could involve taking half of the class and lining them up in the center of the room. With twelve to fifteen students in a single-file line facing the same direction, the teacher asks students to raise their arms above their heads. As the students comply, the teacher explains that their hands indicate sodium ions outside of the neuron's axonal membrane (which they can visibly see as the line of students). The teacher then moves to the back of the line and explains that he or she will tap the last student on the back to initiate action potential (simulating threshold). When the student feels the tap, he or she brings down his or her hands to tap the next student (sodium ions rushing into the axon). Then, the same student immediately snaps up his or her hands and shakes them up and down (sodium-potassium pumps returning to resting potential). When done correctly, this creates a visible wave down the line of students (moving action potential).

In another example, to show the refractory period, the teacher may tap the last student in the line multiple times in succession, but the rate of the waves will not match because of the time it takes each student to shake (time necessary for sodium-potassium pumps to restore resting potential). When half of the students participate in the demonstration, then those students become the audience for their classmates. Body representations tend to be most effective when all students are involved, and the spotlight is not on only a few students.

Element 26: Maintaining a Lively Pace

Pacing refers to running the class in an efficient manner in response to the engagement levels and readiness of learners. Efficiency relates to the speed of the lesson. If the pace of the lesson is too fast, learners will likely get frustrated and struggle. If the pace is too slow, learners may become bored, and off-task behaviors may emerge. A pace that is lively and moving the class along in productive ways is the sweet spot that all science teachers should strive for. This sweet spot might change based on context, and the teacher should monitor the class for cues to speed up or slow down. When teachers maintain a lively pace, they generate heightened energy levels in students.

Figure 7.4 (page 110) depicts the self-rating scale for maintaining a lively pace.

Score	Description
4: Innovating	I engage in all behaviors at the Applying level. In addition, I identify those students who do not exhibit increased energy levels and design alternate activities and strategies that maintain a lively pace in order to meet their specific needs.
3: Applying	I engage in activities to maintain a lively pace without significant errors or omissions and monitor the extent to which students exhibit increased energy levels.
2: Developing	I engage in activities to maintain a lively pace without significant errors or omissions.
1: Beginning	I engage in activities to maintain a lively pace but do so with errors or omissions, such as not slowing down when students are confused and not varying the pace when it is clear that a change of pace would be useful.
0: Not Using	I do not engage in activities to maintain a lively pace.

Figure 7.4: Self-rating scale for element 26—Maintaining a lively pace.

The following strategies illuminate this element.

- Instructional segments
- Motivational hooks

Instructional Segments

To sustain an effective and dynamic science learning environment, teachers must adequately prepare and plan to move students through a variety of experiences. Middle school and high school teachers must proactively think about the length of their class periods (single period, multi-period, block, or other) and consider learning in terms of modules. The amount of time allocated to an activity should not simply be the length of the period (or arbitrary portion of a day) but should fit with the expected learning outcomes. In addition, module lengths should be age- and skills-appropriate. A general rule of thumb supported by attention research is that attention span is double or triple the age in terms of minutes (Brain Balance, 2022).

- **Two years old:** Four to six minutes
- **Four years old:** Eight to twelve minutes
- **Six years old:** Twelve to eighteen minutes
- **Eight years old:** Sixteen to twenty-four minutes
- **Ten years old:** Twenty to thirty minutes
- **Twelve years old:** Twenty-four to thirty-six minutes
- **Fourteen years old:** Twenty-eight to forty-two minutes
- **Sixteen years old:** Thirty-two to forty-eight minutes

Although these are very general guidelines, they are informative reminders to vary activities, transition appropriately, and give periodic breaks, especially if the learning period exceeds one hour. In a science classroom, the time may be broken up into learning segments; for example, modeling activities, laboratory investigations, mini explorations of phenomena (demos), mini-lectures, independent reading time, and a variety of other collaborative learning strategies (including many discussed earlier in this book). Ultimately, teachers must be cognizant of avoiding long periods of teacher-directed instruction (direct lecture) and hold both students and themselves accountable to predetermined timed segments.

To keep segments running smoothly, public timers are a great way to keep everyone on track. There are a variety of timers preprogrammed to specific segment lengths available via a Google search. For instance, if a teacher is planning a fifteen-minute segment, a search for *15-minute timer* will return a variety of possibilities to display in the classroom or simply use as an audible alarm at the end of the defined period.

Motivational Hooks

When beginning an extended exploration, it is often useful to employ motivational hooks to get students excited about the ideas, contexts, or relationships to their lives. Motivational hooks are compelling because they are out of the ordinary, surprising, or counterintuitive. Hooks are also typically short in duration followed by lively discussion or sense-making. In addition, motivational hooks are often delivered through a change-of-pace modality like a read-aloud story, lab demo, video clip, audio clip, or attention-grabbing newspaper headline. The following sections detail two specific approaches: the *con job* and *curiosity creators*.

Con Job

The con-job technique involves the teacher presenting a scenario or phenomenon that isn't what it appears to be. The goal of the con job is to engage students in observations and trick them into making unwarranted inferences. This exposes biases that all observers may bring with them. The con-job technique works well at any level in conjunction with lessons about the scientific process or how to conduct experimental observations.

A classic con job is the candle observation. At the beginning of class, the teacher lights a candle (which is actually an apple core with a shaved pine nut as the wick). As the candle burns, the teacher asks students to write as many quantitative and qualitative observations about the candle as possible within a two-minute time period. After the two-minute period is over, the teacher asks students to share observations and make inferences over a five- to ten-minute timeframe. Finally, the teacher asks students to make quick hypotheses to identity the object. After the class votes, the teacher blows out the candle and then nibbles at a portion of it. The class will be shocked, as students most likely believe this is a candle and not an apple core. When the teacher lit the pine nut, it burned due to the high fat content. This is a great hook and provides a wonderful way to segue into discussions about assumptions and biases in experimentation.

Curiosity Creators

In addition to the con-job activity, curiosity creators and provocative questions are wonderful motivational hooks. Curiosity creators draw student attention based on their novelty (Kise, 2021). For example, the mystery draw is a type of curiosity creator that asks students to draw tasks from a novel source: a decorated container, a fancy bag, pouches affixed to the wall, or other receptacles. The variations on this theme are endless, but each student should at some point select a task from the receptacle, attempt to complete it, and then share it in some way with a small group to achieve a larger understanding. The activity can involve a single container or multiple containers, but the activity works best if every task results in a classwide understanding. Large, mobile whiteboards (huddleboards) may be helpful for students to begin integrating disparate answers or tasks and bring the big ideas into focus.

The mystery draw does not have to involve questions. A simple version of this activity might involve twenty-four different statements written on paper in a single container. Each paper may provide a small piece of necessary information needed to answer a larger question. As students return with their pieces of paper, they realize they must aggregate answers in small groups and form larger groups. An example might be asking the class, "What are the Northern Lights, and how does the phenomenon work?" With the mystery draw, each student gets a piece of the puzzle, and the explanation comes into focus through vigorous expanding student collaborations.

Element 27: Demonstrating Intensity and Enthusiasm

Teacher intensity and enthusiasm about content can be contagious. When teachers share scientific knowledge and phenomena in a way that shows that they are personally connected and excited about the material, students might also share their own passions and interests, allowing them to own their learning. When a teacher is expressive and enthusiastic about the learning, many students will be caught up in the same emotions.

The strategies in this element can help science teachers demonstrate intensity and enthusiasm. Figure 7.5 depicts the self-rating scale for this element.

Score	Description
4: Innovating	I engage in all behaviors at the Applying level. In addition, I identify those students who do not exhibit increased energy levels and design alternate activities and strategies that amplify intensity and enthusiasm to meet their specific needs.
3: Applying	I engage in activities to demonstrate intensity and enthusiasm without significant errors or omissions and monitor the extent to which students exhibit increased energy levels.
2: Developing	I engage in activities to demonstrate intensity and enthusiasm without significant errors or omissions.
1: Beginning	I engage in activities to demonstrate intensity and enthusiasm but do so with errors or omissions, such as not demonstrating intensity and enthusiasm at times when it is clear students need a boost of energy and demonstrating intensity and enthusiasm so frequently that it lessens its effect.
0: Not Using	I do not engage in activities to demonstrate intensity and enthusiasm.

Figure 7.5: Self-rating scale for element 27—Demonstrating intensity and enthusiasm.

Of the strategies listed in appendix A (page 149) for teaching with intensity and enthusiasm, we focus on the following here.

- Nonlinguistic representations
- Personal stories

Nonlinguistic Representations

Nonlinguistic representations are wonderful tools to elicit interest or engage students in an intriguing problem. Earlier in the book, nonlinguistic representations (for example, flow charts, concept webs) served as ways to organize disparate chunks of content (see element 8, page 44). As a part of this element, teachers can use nonlinguistic representations to initiate students in the critical-thinking process. A great example of this is Karl Duncker's (1945) candle problem described in Daniel Pink's *Drive* (2009). The teacher simply shares the image in figure 7.6 and asks students to hypothesize how to use only these three materials to affix the candle to a wall so the wax doesn't drip on the floor. The materials are quite easy to procure, and you may choose simply to have students try to affix the candle to a classroom bulletin board.

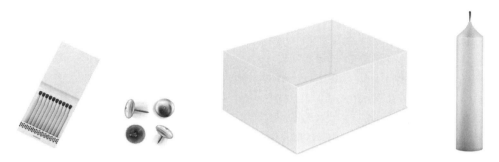

Source: Adapted from Duncker, 1945.

Figure 7.6: Candle problem materials.

Students then share their solutions. After the sharing session, the teacher can use an image to detail the most efficient solution (see figure 7.7).

Source: Adapted from Duncker, 1945.

Figure 7.7: Candle problem solution.

This activity is compelling and surprising because most students won't view the box that contains the tacks as having any functionality. In their minds, it is only used as the storage vessel for the tacks. Few students realize that it can be repurposed. In addition, the activity allows students full autonomy, as its only requirement is that they have to construct a solution based on the image presented to them.

Personal Stories

When teachers tell personal stories in class, they make learning come alive with unique contexts while also providing students with powerful insights regarding their teachers' lives outside the classroom. These stories are especially helpful if the teacher's experience has been documented with photos, videos, journal entries, or other supporting materials. In addition, when teachers are open to answering a variety of questions related to both the scientific content and their personal life histories, discussions are typically lively. When a teacher opens up about his or her personal history, this is an invitation to students to share their own. This sharing is also a great time to highlight personal backgrounds and celebrate the diversity of student experiences.

For example, in a high school chemistry classroom, author Brett Erdmann incorporated personal story-telling in the context of a nuclear chemistry unit. The author had visited Hiroshima while living in Japan during a semester abroad. After discussing the chemistry behind nuclear weapons and nuclear reactors, Brett displayed a slide show of the photos he had taken at the Hiroshima Peace Memorial Museum. In the context of discussing the photos, Brett was able to describe the stories behind the images. In addition to sharing photos, Brett had also kept a journal and shared snippets of various entries. Students were equally fascinated by the nuclear science, World War II politics, and college-age experiences of their teacher. With a significant number of Asian students in the class, respectful discussion of the differing viewpoints ensued regarding the necessity of dropping the bomb (and students were able view it through both American and Japanese lenses). In this example, personal storytelling provided a wonderful context for interdisciplinary learning and the sharing of lived experiences.

Element 28: Presenting Unusual Information

The history of science and scientific discovery is filled with unique, unexpected, and awe-inspiring work. In addition, the scientists, historical contexts, and continuously changing nature of scientific theory provide fertile ground for discussions that connect to students' lives. Unusual information generally stimulates students' interest and intrigue and is an excellent way to provoke engagement.

This section discusses strategies for presenting unusual information in science. Figure 7.8 depicts the self-rating scale for element 28.

Score	Description
4: Innovating	I engage in all behaviors at the Applying level. In addition, I identify those students who do not exhibit increased interest and intrigue in the content I present in class and design alternate activities and strategies that mention unusual information to meet their specific needs.
3: Applying	I engage in activities to present unusual information without significant errors or omissions and monitor the extent to which students exhibit increased interest and intrigue in the content I present in class.
2: Developing	I engage in activities to present unusual information without significant errors or omissions.
1: Beginning	I engage in activities to present unusual information but do so with errors or omissions, such as presenting unusual information that has little or no relationship to content I present in class and not allowing adequate time for students to discuss and react to the unusual information.
0: Not Using	I do not engage in activities to present unusual information.

Figure 7.8: Self-rating scale for element 28—Presenting unusual information.

The following strategies apply to this element.

- Fast facts
- History files
- Guest speakers

Fast Facts

Focusing on unusual facts (or sharing facts in unusual ways) helps students engage their creative impulses and taps into the joy of sharing secret knowledge. Fast facts are truly engaging and leave the recipient wanting to know more. A fast fact should be novel and possibly a bit unexpected. For instance, a fast fact regarding dolphins is that they can breathe underwater for eight to ten minutes. The fact begets questions like "How?" or statements like "Wow, that is a long time!" For a fast-facts activity to be effective, teachers should provide students with a template, allow time for research, and create a fun atmosphere for sharing. A template may involve the fact and possibly an accompanying interesting image, followed by a space for students to ask questions. Once students record their questions, they can cite evidence to help answer each question. Following this process, the teacher may lead a whole-class discussion that tries to explain how this phenomenon is possible.

In a biology class, author Brett Erdmann annually conducts an "organelle speed dating" activity. He asks students to each create a dating profile for a specific organelle using a presentation slide template (such as Google Slides or Microsoft PowerPoint). Once each student conducts the research and populates the profile, the teacher sets a timer for three minutes and has pairs of students (who each have a different organelle) share their dating profiles.

After each three-minute segment, the teacher rotates the students to different pairs. In the background, pop love songs play softly. After a number of rotations, students have shared organelle information in unique ways as well as unexpected facts. Figure 7.9 shows an example of a teacher-created slide.

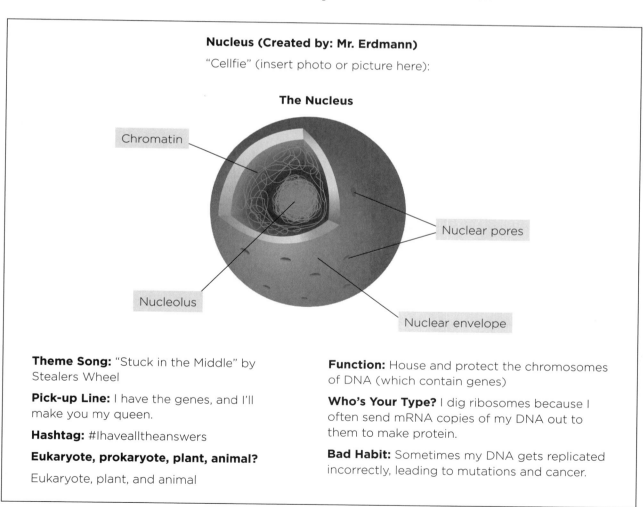

Figure 7.9: Teacher-created slide.

History Files

History files or historical perspective analysis generally involves the use of primary source materials and requires students to read them, construct meaning from the information, and place them in proper historical context. History files or historical perspective analysis may initially seem out of place in the science classroom. In specific situations, they can help students understand a timeline of scientific discovery or the shifting understandings that coalesced into modern theories. This approach has the potential to bring relevance to scientific content by exploring the importance of the people, places, and events that shape ideas.

One example of a historical approach is an analysis of the scientific work leading to the discovery of DNA molecules as genetic material. In a biology class, a teacher could assign student groups to work on one of the following scientists or scientist pairs: Frederick Griffith, Oswald Avery, Alfred Hershey and Martha Chase, Erwin Chargaff, Rosalind Franklin, or James Watson and Francis Crick. Students then research the work of these scientists and complete the activity shown in figure 7.10. Since this work can be involved and may require periodic teacher guidance, student work should be completed in class.

The History of the Discovery of DNA as the Genetic Material ("Tweets" Activity)

As an individual, it is your job to complete the following task.

  1. For homework, read the selected article for **your** scientists, and answer the reading questions.

As a group, it is your job to complete the following tasks.

  2. Share individually constructed answers orally with the group. Agree on the answers for the reading questions about your scientists.

  3. Create **six presentation slides** that contains the following:
     a. Title slide with scientist's names, photos of scientists, class period, and learner names
     b. Five slides, each containing an **image** and a **tweet**.

  Find five separate images from the internet that would be *most* useful in helping explain the work of the scientists.

  In addition, write five tweets (140 characters or less) that explain the concepts that go along with each image.

  4. Create **six multiple-choice questions** (four choices for each) that quiz essential aspects of the work of the scientists. Use the questions that you answered from step 2 as a reference. Also, use the original reading (from step 1) and book reading as guides. Collate and edit all questions in a shared group document. Make sure you have a .jpg image associated with each question.

  5. Each group presents its slideshow. Each group stands in front of the class as coteachers and leads learners (your classmates) through the activity. Be sure to stop after each slide and examine the data. Provide feedback to the class and elaborate on answers when necessary.

Figure 7.10: History files activity example—Discovery of DNA.

Further, while sharing the information, students can jigsaw the history into a comprehensible sequencing of the discovery process with their own note-taking template, as shown in figure 7.11.

Guest Speakers

Guest speakers inject much-needed relevancy and context into class lessons. They also are often easy to locate within your own community. Many student parents or guardians with expertise in a science or curriculum-related field are happy to commit to a day of learning and share their educational and career experiences.

	Griffith	Avery	Hershey and Chase
Organism or Material			
Experiment or Method			
Conclusion			

	Chargaff	Franklin	Watson and Crick
Organism or Material			
Experiment or Method			
Conclusion			

Figure 7.11: Student note-taking template.

In addition, parents or guardians may have very specialized skill sets that translate to wonderfully challenging and engaging lessons. Author Brett Erdmann recently invited a parent who was a geneticist to teach part of a class. The parent led students through a wonderfully complex pedigree based on a real-life clinical scenario. In addition to this activity, students were able to interview the guest about medical school and the day-to-day experience of a clinical researcher. Brett also integrated a student-written blog project into his class, involving the long-term reading of a popular science nonfiction book. In conjunction with the blog projects, he contacts the authors of these *New York Times* bestselling books and asks if they would be available to speak with his classes. With just a simple email request, four world-renowned scientist authors have either visited his classes or connected with students via video conferencing.

With Zoom and similar platforms being so prevalent and accessible to educators and industry professionals alike, the possibility of accessing a significant scientist or academic in your classroom is greater than ever before. Don't be afraid to ask anyone to come to your classroom. You might be surprised by who says *yes*. Authors Steve Wood and Brett Erdmann have personally witnessed how many scientists and STEM professionals value the work of educators, have a sense of gratitude, and wish to give back.

Element 29: Using Friendly Controversy

Students may become more engaged when exploring controversial topics. To help stimulate, intrigue, and enhance subject-matter knowledge, teachers can employ strategies that foster controversy (or disagreement) in a civil manner. When executed well, friendly controversy helps students analyze content with a critical eye. Scientific discoveries have often been at the center of controversies, and many scientists were initially vilified before their work was accepted by the scientific community. Controversy continues in science in the present day. Many ideas are being hotly debated that may have an impact on human health, the environment, and technological innovation. In the classroom, teachers can explore these controversies in a way that provides psychological safety, honors student voices, and models the tenets of respectful debate.

This section presents strategies for using friendly controversy in science. Figure 7.12 depicts the self-rating scale for this element.

Score	Description
4: Innovating	I engage in all behaviors at the Applying level. In addition, I identify those students who do not exhibit increased interest and intrigue in the content I present in class and design alternate activities and strategies that use friendly controversy to meet their specific needs.
3: Applying	Along with adequate guidance and support, I engage students in activities that involve friendly controversy without significant errors or omissions and monitor the extent to which students exhibit increased interest and intrigue in the content I present in class.
2: Developing	I engage students in activities that involve friendly controversy and provide adequate guidance and support.
1: Beginning	I engage students in activities that involve friendly controversy but do not provide adequate guidance and support, such as structuring the activity so students have clear roles and responsibilities and not allowing the controversy to become emotionally charged.
0: Not Using	I do not engage students in activities that involve friendly controversy.

Figure 7.12: Self-rating scale for element 29—Using friendly controversy.

We recommend the following strategies for using friendly controversy in science.

- Seminars
- Diagrams comparing perspectives

Seminars

Seminars are safe spaces that are organized thoughtfully to facilitate large-group discussions. In a science classroom, it's often helpful to examine multiple viewpoints in safe and nonthreatening spaces prior to diving deep into scientific details. Seminars work well for having students examine scientifically supported ideas in a scaffolded way while using high-quality supporting resources. In small groups, teachers direct students to access resources, assimilate the findings, and agree on a position for their group. Groups then share their positions with the whole class.

Seminars are often organized around an effective essential question. In a biology classroom, a question related to an ecological unit might be: *If the entire human population adopted a vegetarian diet, what would happen to the biosphere?* The teacher could give students think time and ask them to individually address at least three of the ideas (from past and current units) shown in figure 7.13.

Food Pyramids	Biogeochemical Cycles
Food Webs	Population Dynamics and Limiting Factors
First Law of Thermodynamics	Photosynthesis and Cell Respiration
Levels of Ecological Organization	Others? You may choose anything specific to a past learning target.

Figure 7.13: Three ideas related to essential question about vegetarian diet.

After developing individual written responses, students gather in groups to identify the three most compelling pros and three most compelling cons from the article "Should People Become Vegetarian?" (ProCon, 2021; http://vegetarian.procon.org) and add them to a shared whiteboard (huddleboard). Following this step, students add their individual thoughts to construct a claim for or against vegetarianism based on the information gathered in the individual and group activities. Finally, each group must come to a consensus and make a claim for or against a vegetarian diet from an *ecological* standpoint through the production of three memes.

Diagrams Comparing Perspectives

Students can show varying points of view with Venn diagrams, which highlight areas of congruence and areas of disagreement between two or more ideas or concepts (see element 10, page 59). The diagrams typically work very well with related theories. For example, historically, there have been many theories describing the *how* behind evolutionary change. The Venn diagrams in figure 7.14 (page 120) can be useful in clarifying similarities and differences between these theories.

Other interesting examples explore the different lenses (scientific or socially constructed) used to examine a central theme. For example, a biology class might explore the relationships between biological sex and gender (figure 7.15, page 120).

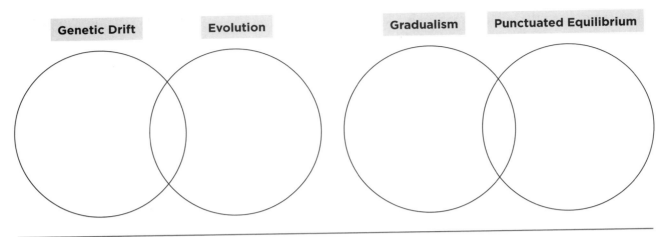

Figure 7.14: Venn diagrams to identify similarities and differences.

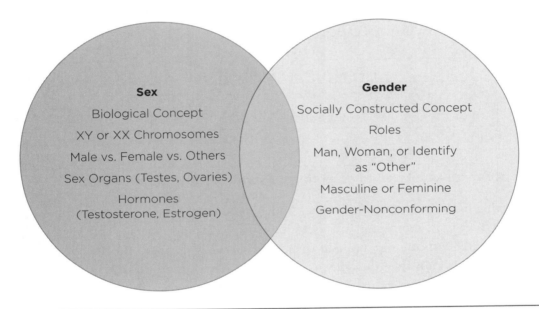

Figure 7.15: Venn diagram comparing biological sex and gender.

By using a Venn diagram like the one in figure 7.15, the teacher may direct a discussion about the overlap between sex and gender. With the outer circles prepopulated, students can use that information to fill in the overlapping middle section (perhaps supported by supplementary reading texts). With the sex and gender example, interesting discussions can ensue regarding the pitfalls of conflating these ideas and the impact of misunderstandings on society.

As students create their diagrams, encourage them to research and find evidence to support the major claim or idea in each circle. Venn diagrams can become more complex as students progress through a learning sequence. The diagrams may be altered to show the intersectionality of three or more ideas. Venn diagrams also can be modified to form more elaborate graphic organizers or expanded to form complex concept webs. Science teachers should encourage students to move beyond the basic form by incorporating graphical elements that extend the claims and make new connections.

Element 30: Using Academic Games

Academic games are quick remedies for disengagement and provide students with a fresh look at content. Although academic games are inherently fun, the purpose of these games should be to encourage higher-order thinking. The games should encourage students to make content connections, evaluate evidence, judge models, compare and contrast, and transfer knowledge to novel scenarios. Although games can be competitive in nature, teachers must be mindful to ensure *all* students can interact, not just the fastest or highest-achieving students.

The following element presents strategies for using academic games in science. Figure 7.16 depicts the self-rating scale for this element.

Score	Description
4: Innovating	I engage in all behaviors at the Applying level. In addition, I identify those students who do not exhibit increased interest and intrigue in the content I present in class and design alternate activities and strategies to meet their specific needs.
3: Applying	Along with adequate guidance and support, I engage students in activities that involve academic games and monitor the extent to which students exhibit increased interest and intrigue in the content I present in class.
2: Developing	I engage students in activities that involve academic games and provide adequate guidance and support.
1: Beginning	I engage students in activities that involve academic games but do not provide adequate guidance and support, such as establishing clear roles and procedures for students and involving inconsequential competition.
0: Not Using	I do not engage students in academic games.

Figure 7.16: Self-rating scale for element 30—Using academic games.

Teachers can consider the following strategies for incorporating academic games into instruction.

- Which one doesn't belong?
- Inconsequential competition
- Vocabulary review games

Which One Doesn't Belong?

In this game, teachers create four clues—three that are correct and one that is incorrect. Sometimes this game is referred to as *three truths and a lie*. Use this game to help students develop scientific vocabulary, discriminate process steps, learn specific content information, or expose misconceptions. Students read the four clues, determine the incorrect choice, and orally (or in writing) explain why that choice doesn't fit.

Figure 7.17 (page 122) shows an example from a freshman-level biology class studying photosynthesis. The number of choices can vary, and teachers may adjust them based on students' developmental levels and the complexity of the concepts. A fun variation on the game involves asking each student to create the truths and lie for a specific concept on a notecard and then use it to try to trick or stump other students working in small groups.

Which One Doesn't Belong?
Term: Chloroplast
1. Double-membraned organelle containing the pigment chlorophyll 2. Evolved from prokaryotes according to the endosymbiotic theory 3. Present in both eukaryotes and prokaryotes 4. Site of chemiosmosis and ATP production
Term: Oxidation
1. Molecule gains electrons 2. Molecule changes to have a lower potential energy 3. Always happens in association with a reduction 4. Does not have to directly involve oxygen
Process: Light Reactions
1. Occur along the thylakoid membrane of the chloroplast 2. Lead to the production of ATP and NADPH 3. Chlorophyll pigments are oxidized by the sun's photon energy 4. Oxidize ATP and NADPH
Process: Calvin Cycle
1. Fixes carbon dioxide gas to form glucose 2. Reduces NADPH and ATP 3. Occurs in the stroma of the chloroplast 4. Catalyzed by a series of enzymes
Process: Carbon Cycle
1. Products from photosynthesis become reactants for cell respiration. 2. Photosynthesis removes carbon dioxide from the air. 3. Cellular respiration breaks down glucose to release carbon dioxide gas. 4. The burning of fossil fuels removes carbon dioxide gas from the air.

Figure 7.17: Which one doesn't belong? example.

Inconsequential Competition

Competitions between individuals or small groups are inherently engaging for students. As long as the competitions don't impact scoring or grading (they are inconsequential), teachers may find that some students are motivated by the lively pace, unique framing of questions, and rapid feedback. In addition, prizes like stickers, candy, and other small items are appealing to students of all ages. Many free online technology tools have been introduced since 2010 that are helpful for teachers attempting to run in-class competitions; we cover two such tools in the following sections, Quizizz and Kahoot! These tools are especially engaging if students have their own devices (such as tablets, smartphones, or laptops).

In addition to online games, paper and whiteboard games are equally effective for fostering inconsequential competition. Teachers can display a *Jeopardy!*-style game with categories and point totals on a classroom chalkboard or whiteboard. In this scenario, teachers divide the room into teams. Each team gets a buzzer, bell, or other noise-making device.

For each round of questions, the team selects a new team member. The teacher then asks the newly selected team members from each team to pick both the category and point amount. Then the teacher presents the questions and waits for one of the students to "buzz in" (use their noisemaker to indicate that they wish to answer the question). With a correct answer, the teacher records the points on the class whiteboard.

This process repeats until all questions have been asked and all point amounts are off the board. If the classroom is a typical size, all students will be able to answer at least one question. If the classroom is larger, the teacher may run another round of *Jeopardy!* with a new gameboard. An example of a *Jeopardy!*-style gameboard from a formula-writing and reactions chemistry unit is shown in figure 7.18.

Elements and Compounds	Names to Formulas to Names	Single and Double Displacement	Decomposition, Combustion, and Synthesis	Science Grab Bag
100	100	100	100	100
200	200	200	200	200
300	300	300	300	300
400	400	400	400	400
500	500	500	500	500

Figure 7.18: *Jeopardy!*-style gameboard.

Quizizz

Quizizz (quizizz.com) provides an online platform for teachers to incorporate multiple-choice questions and optional embedded graphics. When a teacher launches the competition, students answer each question as an embedded timer ticks down. If a student answers quickly, they receive more points. As all students progress through the questions, they receive updates about their ranking based on their current point totals. In addition, each correct answer and incorrect answer is followed by a humorous meme that keeps students moving. At the end of the competition, students receive a report of all the questions with correct and incorrect answers. Throughout the game, the teacher does not need to moderate. In fact, as the game proceeds, the teacher gets real-time information regarding every student's progress (percent correct and points) and the overall running percentage correct. The teacher may choose to review most-missed questions at the end of the game.

Kahoot!

Kahoot! (Kahoot.com) is another online platform that requires teacher moderation. The teacher preloads a series of multiple-choice questions (with optional embedded images) and then reveals each question on a projector screen; a timer counts down with each question. Connected student devices display shapes or certain colors that correspond to the answers on the screen, and students choose what they believe are the correct answers. When time is called, the screen shows which answers were selected and how many students selected each answer. The teacher may pause after each question to review the right answer and talk about misconceptions or proceed to the next question without elaboration if all students were correct. At the end of the game, a leaderboard is displayed for first, second, and third place. At this point, the teacher may choose to recognize the leaders with prizes or low-stakes accolades.

Vocabulary Review Games

Success in many science courses relies on facility with vocabulary that is unfamiliar to students. In some ways, learning science can be a lot like learning a foreign language. To help students build vocabulary, teachers may wish to incorporate various forms of card games. A very simple version of this game involves having students create notecards for essential vocabulary (one side with the term and one side with the definition). After each student creates his or her own stack, the teacher places students in pairs or small groups to quiz each other and record the number of correct answers. Typically, using a group timer set at five to ten minutes keeps the action lively, and students speed through the cards as they try to accumulate the most points. In addition, the teacher may instruct students to put incorrect cards in a separate stack to revisit later in the game (or in a separate speed round).

For teachers who prefer to use online tools with students who have their own internet-connected devices, Quizlet (Quizlet.com) has a very compelling Quizlet Live feature in which an existing online card deck of vocabulary can be instantly transformed into an online game. After students log in on their device (for example, a tablet, phone, or laptop), the platform randomly assigns them to teams. Students then move to tables with their assigned groups, and the game begins. Within their groups, the platform shares the term and offers different choices to each student in the group. The group must collaborate, share screen information, and decide which student has the correct choice. If the correct choice is made, the group gets the next term. Note that each group receives different cards, so communication between groups does not help other groups succeed. The goal is to get twelve correct in a row. If any group gets one term wrong, group members have to start over and continue a new string of correct answers. The first group to get twelve in a row wins. Typically, the deck has at least thirty to forty terms so the teacher can run the game two or three times without students getting repeats.

Element 31: Providing Opportunities for Students to Talk About Themselves

When students have opportunities to talk about themselves, they feel more welcome in the learning environment (Marzano, 2017). By personalizing the learning experience in this way, we hope to create learners who can see themselves as capable scientists, perhaps envisioning viable pathways to science and STEM careers. Scientific information is sometimes considered inaccessible by the general public, and this has numerous societal implications (including a growing anti-science movement). When students bring their lives into the scientific work, real-world relevance is often a natural byproduct. Drawing on students' cultures and geographic locations to engage them in learning allows students to equitably engage in authentic phenomena, one of the overriding goals of the NGSS (Gewertz, 2020).

In this section, we share strategies teachers can use to provide opportunities for students to talk about themselves in the context of scientific discovery and science content discussions. Figure 7.19 depicts the self-rating scale for this element.

Teachers can consider using the following strategies to help incorporate opportunities for students to talk about themselves.

- Interest surveys
- Student learning profiles

Score	Description
4: Innovating	I engage in all behaviors at the Applying level. In addition, I identify those students who do not appear to be motivated to participate in classroom activities and design alternate activities and strategies for students to talk about themselves in ways that meet their specific needs.
3: Applying	I engage in activities that provide students with opportunities to talk about themselves without significant errors or omissions and monitor the extent to which students appear to be motivated to participate in classroom activities.
2: Developing	I engage in activities that provide students with opportunities to talk about themselves without significant errors or omissions.
1: Beginning	I engage in activities that provide students with opportunities to talk about themselves but do so with errors or omissions, such as not making links with the content I present in class and not providing adequate time for students to talk about themselves.
0: Not Using	I do not engage in activities that provide students with opportunities to talk about themselves.

Figure 7.19: Self-rating scale for element 31—Providing opportunities for students to talk about themselves.

Interest Surveys

Interest surveys are effective tools to gather information from students regarding learning goals, learning preferences, personal and family histories, class expectations, personal interests and hobbies, and a variety of other useful data. Teachers collect student information using low-tech paper surveys that they compile and keep in a file. They might also create electronic forms to keep a variety of information to access and sort later in the year for specific purposes. For example, a teacher might use a Google form to collect student birthdays, which they can sort sequentially to recognize students throughout the year.

Although interest survey information is especially instructive to the teacher when attempting to differentiate in the science classroom, the information may be equally important to share with the class to build a strong learning culture. For example, an All About Me project can consist of a visual show and tell using a presentation tool like Google Slides or Apple Keynote. The teacher asks each student to add photos to a digital bulletin board. The criteria for the photos can be modified for any classroom. Students are usually eager to share photos similar to those they share on social media platforms. Once each student creates a digital bulletin board, the teacher allows a few minutes at the beginning of each class for one or two students to share their boards. Sharing can continue over many weeks until all students have shared.

To begin this activity, the teacher might share his or her own All About Me board so students get to know their teacher as a person and realize that he or she is learning along with them. The activity works especially well if the teacher encourages the class to ask questions to extend the conversation beyond talking about photos. Figure 7.20 (page 126) shows a directions template for the activity, along with a teacher-created example.

Student Learning Profiles

To develop student learning profiles, teachers may use a survey to collect information from students regarding their perceptions about how they learn best, or their most successful learning strategies. Learning profiles also may capture how students enjoy expressing themselves.

ALL ABOUT ME—DIGITAL BULLETIN BOARD

Design your bulletin board to help me get to know you better. Include **at least 10** (or more!) things that tell me more about who you are.

Here are some ideas of what to include:

- REQUIRED: Your name (first and last)
- REQUIRED: A picture of you (Selfies are great!)
- Other ideas:
  + Preferred pronoun
  + Pictures of family, friends, and pets
  + Your favorite book(s)
  + Your favorite movie(s)
  + Your favorite TV show(s)
  + Your favorite music

More Ideas:
- What you like to do for fun
- Ethnic, religious, and/or cultural background
- A future goal you have
- Your favorite sport and/ or team
- Your favorite food
- Your favorite subject to learn about
- A hobby you enjoy
- A job you have or have had
- A college or career you are interested in entering
- A place you'd like to visit
- Something you did over the summer
- Your favorite quote
- Be creative!

Happiness is the only good. The time to be happy is now. The place to be happy is here. The way to be happy is to make others so...

Robert G. Ingersoll

Source: BrainyQuote, n.d.

Figure 7.20: All About Me activity example.

Teachers may choose to organize student responses into categories of cognitive processing styles, as described in the book *Doable Differentiation* (Kise, 2021): (1) structure and certainty, (2) experience and movement, (3) vision and interpretation, and (4) question and communication. These four styles are summarized by the following: "let me know what to do," "let me do something," "let me follow my own lead," and "let me lead as I learn" (p. 18).

After categorizing responses under these processing styles, the teacher may use the information to ensure a variety of choice in projects or intentionally stretch students to engage in styles outside of comfort zones indicated in the survey. Focusing on cognitive processing styles reminds teachers to include a variety of activities to appeal to *all* learners. Teachers should not categorize students by any particular learning style, but instead should stretch themselves over the course of the year to engage with all styles.

Element 32: Motivating and Inspiring Students

Ultimately, the highest form of engagement occurs when students are intrinsically motivated, leading to student success and well-being (Howard, Bureau, Guay, Chong, & Ryan, 2021). Intrinsic motivation often develops by giving students opportunities to self-actualize and connect to something greater than themselves (Marzano, 2017). The process of developing motivated students can be challenging, and teacher persistence is required. Since science classes generally become more vocabulary-rich and conceptually complex through the high school grade levels, some students lose the motivation to continue with science coursework. Science teachers should keep in mind that almost all kindergartners are natural scientists who are excited to ask questions and figure things out. Science teachers can employ specific strategies to recapture this spark of wonder and develop students who employ this approach joyfully.

This section describes motivational and inspirational strategies for science teachers. Figure 7.21 depicts the self-rating scale for this element.

Score	Description
4: Innovating	I engage in all behaviors at the Applying level. In addition, I identify those students who do not appear to be developing a sense of self-agency and design alternate activities and strategies to meet their specific motivational needs.
3: Applying	I engage in activities to motivate and inspire students without significant errors or omissions and monitor the extent to which students appear to be developing a sense of self-agency.
2: Developing	I engage in activities to motivate and inspire students without significant errors or omissions.
1: Beginning	I engage in activities to motivate and inspire students but do so with errors or omissions, such as not allowing enough time for the activities and not communicating the importance and relevance of these activities to students.
0: Not Using	I do not engage in activities to motivate and inspire students.

Figure 7.21: Self-rating scale for element 32—Motivating and inspiring students.

In appendix A (page 149), we list various strategies for motivating and inspiring students. Here we highlight the following specific strategies.

- Possible selves activities
- Gratitude journals

Possible Selves Activities

Although not all students will become doctors or science researchers, teachers should provide many opportunities for students to envision themselves working in a STEM (science, technology, engineering, mathematics) field. For example, after taking a science course or learning science in elementary grades, students should know what it feels like to conduct a successful controlled experiment. As discussed earlier in the chapter, guest speakers can help students envision themselves as future STEM professionals. In short, students should have a good sense of how the curriculum relates to both their current and future lives. Students can extend and apply learning outside the classroom through local job shadowing, internships, and adult mentorships (often in conjunction with science fair or independent study projects).

Gratitude Journals

In his book *The Happiness Advantage*, researcher Shawn Achor (2010) states, "When we are happy—when our mindset and mood are positive—we are smarter, more motivated, and thus more successful. Happiness is the center, and success revolves around it" (p. 37). Achor claims that the learning centers of the brain are more effective when the mood is positive as opposed to neutral or stressed. He also states that positivity in the moment leads to a variety of successful outcomes in education, business, and medicine. In short, happiness (positivity) leads to success and not vice versa (Achor, 2010).

Robert A. Emmons and Michael E. McCullough (2003) state that if people consistently maintain a gratitude journal where they record three positive things daily, they can develop a mindset that scans the world preferentially for the positive instead of the negative. Teachers and students can apply gratitude journals across all areas of students' lives. Teachers also may choose to periodically direct gratitude reflections toward various aspects of the science learning process.

Gratitude journals can be organized in a variety of ways. A journal can be open-ended, with a simple requirement of having students write one thing they're grateful for on a specific day. Teachers also might use public gratitude walls. A public gratitude wall is a simple idea in which a teacher may reserve space in the classroom and label it with a prompt or stem. The teacher may provide one prompt for the year and ask students to add to it daily. Another option is to provide a new prompt every day and ask students to respond on the classroom chalkboard or whiteboard. Public prompts may begin with various things students appreciate like favorite foods, pets, vacation spots, lab activities, and so on. After students' public postings, the teacher provides stems for students to go deeper in a private space (for example, written journal, digital document). Stems might include the following:

- Why are you grateful for . . . ?
- How would you describe . . . ?
- How will you show . . . ?

Through the gratitude journal process, students identify, reflect, and write, practicing skills that will serve them well in any future scientific investigations.

GUIDING QUESTIONS FOR CURRICULUM DESIGN

This design question focuses on engagement: *What engagement strategies will I use to help students pay attention and be energized, intrigued, and inspired?* The following questions, which are aligned to each of the elements in this chapter, guide teachers to plan instruction.

- **Element 23:** What will I do to notice when students are not engaged, and how will I react?

_____

_____

_____

_____

_____

- **Element 24:** What will I do to increase students' response rates?

_____

_____

_____

_____

_____

- **Element 25:** What will I do to increase students' physical movement?

_____

_____

_____

_____

_____

- **Element 26:** What will I do to maintain a lively pace?

_____

_____

_____

_____

_____

- **Element 27:** What will I do to demonstrate intensity and enthusiasm?

_____

_____

_____

_____

_____

- **Element 28:** What will I do to present unusual information?

_____

_____

_____

_____

_____

- **Element 29:** What will I do to engage students in friendly controversy?

  _____
  _____
  _____
  _____
  _____

- **Element 30:** What will I do to engage students in academic games?

  _____
  _____
  _____
  _____
  _____

- **Element 31:** What will I do to provide opportunities for students to talk about themselves?

  _____
  _____
  _____
  _____
  _____

- **Element 32:** What will I do to motivate and inspire students?

  _____
  _____
  _____
  _____
  _____

Summary

Engaging students in the science classroom encompasses various levels from basic to highest degrees: getting students' attention, producing energy, stimulating intrigue, and motivating and inspiring students intrinsically. Several elements and embedded strategies help teachers design and conduct lessons that foster engagement. They clearly would not use all elements in a single unit of instruction but would frequently—if not daily—notice when students are disengaged or unresponsive. When this occurs, teachers can increase response rates, use physical movement, maintain a lively pace, and demonstrate intensity and enthusiasm. Organizing the classroom layout and understanding students' backgrounds and interests contribute to some aspects of engagement.

CHAPTER 8

Implementing Rules and Procedures and Building Relationships

As we mention in the introduction (page 1), *The New Art and Science of Teaching* framework features three overarching categories (feedback, content, and context), ten teacher actions, forty-three elements, and more than 330 accompanying strategies. Teachers intentionally select elements and embedded strategies to build a well-rounded, effective instructional program based on a unit's learning goals of what students should know, understand, and do. Since our focus in this text is on science learning, this chapter highlights only one of the five elements within the category of implementing rules and procedures and one element within the category of building relationships.

In science classrooms, the management of physical spaces and laboratory equipment is essential to meaningful learning experiences. Schools vary greatly in their ability to build footprints and resources, so it's important that teachers share all strategies, regardless of their school's advantages or limitations. Many prior elements have discussed the importance of motivation and personalizing learning. These elements allow for the opportunity to explore those concepts in detail and make explicit connections to science. In this chapter, we discuss the following elements:

- **Element 34:** Organizing the physical layout of the classroom
- **Element 39:** Understanding students' backgrounds and interests

We encourage readers to study *The New Art and Science of Teaching* (Marzano, 2017) to take advantage of learning about all the strategies related to each category so they are fully aware of what contributes to excellence in teaching.

Element 34: Organizing the Physical Layout of the Classroom

The physical environment of the classroom can help stimulate or hinder students' sense of order. In a science classroom, students need to know how to access materials and manipulate the furniture appropriately to meet the needs of group and laboratory work. The teacher must ensure that classroom space is flexible and can be reconfigured for a variety of collaborative purposes. In addition, classroom space should showcase student work, serve as inspiration to explore future ideas, and be both physically and emotionally safe.

In this element, strategies are shared for organizing the layout of science classrooms and laboratory spaces. Figure 8.1 depicts the self-rating scale for this element.

Score	Description
4: Innovating	I engage in all behaviors at the Applying level. In addition, I identify those students who do not appear to enjoy and use the resources within the classroom and design alternate activities and strategies to organize the physical layout of the classroom to meet their specific needs.
3: Applying	I engage in activities to make the physical layout of the classroom appealing and supportive of learning without errors or omissions and monitor the extent to which students both enjoy and use the resources within the classroom.
2: Developing	I engage in activities to make the physical layout of the classroom appealing and supportive of learning without errors or omissions.
1: Beginning	I engage in activities to make the physical layout of the classroom appealing and supportive of learning but do so with errors or omissions, such as not displaying enough student work and not referencing the displayed work during class time.
0: Not Using	I do not engage in activities to make the physical layout of the classroom appealing and supportive of learning.

Figure 8.1: Self-rating scale for element 34—Organizing the physical layout of the classroom.

This element is associated with the following strategies, which address how organizing the physical layout of the classroom can help students learn science.

- Displaying student work
- Considering classroom materials, computers, and technology equipment
- Placing student desks and planning areas for group work

Displaying Student Work

When teachers showcase student work in the classroom or hallways, students receive the implicit message that their work has value. It also implies student ownership in the learning process. Student displays may be long-term (bulletin boards where the examples remain for weeks or months) or short-term (whiteboard examples posted around the room for a gallery walk within the confines of a single class period). Regardless of the time frame, each piece of displayed work is a snapshot of the learning process and can be incorporated into future instruction. Selection of displayed work can be global (all students) or specific to certain criteria for high quality (the use of exemplars). When displaying work publicly, it is imperative that teachers tell students in advance so they are not caught off guard. Also, if selecting work based on quality, the teacher should find ways to highlight *every student* at least once over the course of the semester (or quarter) to build a culture of pride in good work.

In elementary classrooms, a variety of student displays can enhance science learning. For example, a kindergarten class may display plants being grown from seeds along the ledge near the windows, with pots labeled with students' names. Each student may be charged with the care of his or her plant and guided to communicate with other students about how well plants are growing. A fifth-grade science class might incorporate a research poster project in which they describe the various iterations of devices they constructed in an egg-drop competition and which device was most successful in protecting the egg. Teachers can hang posters in the classroom or outside the classroom in the hallways for all passersby to view.

In later grades, student displays may become more elaborate and involve extensive use of a variety of technologies. A high school biology display might involve posting a time-lapse video on the internet showing a clay model created by a small group of students to explain the process of DNA replication. An AP chemistry display might involve a TikTok video showing a successful titration with embedded text explanations and hyperlinks. Some tools useful to facilitate these displays of student learning include the following.

- **Stop Motion Studio (www.cateater.com):** Use this resource to create time-lapse videos of clay models.
- **TikTok (www.tiktok.com):** Use this resource to create short videos with embedded text to explain a concept.
- **Flipgrid (www.flipgrid.com):** Use this resource to engage in two-way video discussions.
- **Loom (www.loom.com):** Use this resource to create videos overlaid onto slide shows.

In any grade, displays may be in the form of actively running experiments, posters, small drawings, posted videos, or a variety of other visualizations. Note that displayed work does not necessarily have to be completed work. Often, it is helpful to display work that is partially finished and have student groups evaluate the work and provide feedback. Students then use this feedback to make revisions.

Student displays also can contribute to group learning throughout the unit or learning sequence. For example, vocabulary terms are critical for learning a variety of science concepts. At the beginning of the unit, the teacher might choose to create a word wall for all the unfamiliar terms. Each student then defines a single term on an 8.5"× 11" paper. Students also draw a graphic to support the definition of the term that the teacher then posts on the wall, and the wall serves as a visual and textual support system for learning during the unit. When struggling with terms or concepts, the teacher can point to the terms on the wall as reminders to help coach students through scientific explanations that rely on the proper use of vocabulary terms.

Considering Classroom Materials, Computers, and Technology Equipment

When students engage in scientific inquiry and modeling, it is helpful for teachers to provide materials in containers that are easily accessible. Materials may be kept in a bin or basket in the center of a group or lab table. These bins may contain colored pencils, dry-erase markers, erasers, pens, pencils, and other frequently used supplies. If the tables ever need to be cleared quickly for a lab or other activity, the bins may be easily relocated to a different area of the classroom.

During specific laboratory or investigating scenarios, it's often helpful to have all supplies accessible from a central table in the classroom. We jokingly refer to this as *the buffet of learning*. The central location of supplies increases efficiency and works best when it is equidistant from the other lab benches or group tables. In addition, if any of the materials involve a specific safety concern (for example, strong acids in a chemistry classroom or cutting tools for an engineering project), the teacher can manage these items from one location while within easy view of all the lab or group tables.

A well-stocked and organized classroom is essential for students to efficiently engage in effective learning strategies. Large whiteboards (huddleboards) and smaller whiteboards are low-cost materials that can be used for many of the strategies discussed in this book. Huddleboards often come with their own rolling cart that makes for efficient distribution. Small whiteboards may be kept in stacks on a countertop or in a laboratory cabinet or drawer. Labeling all the cabinets and drawers with student-accessible supplies opens up numerous possibilities for students to employ their creativity in various projects. Some of these less frequently used (but important) supplies might include: construction paper, modeling clay, pop beads, card stock, notecards, pipe cleaners, and a variety of other craft or modeling supplies. It's important to keep these areas organized and coach students about how to access these areas, clean up after themselves, and return supplies in their original conditions to the proper locations. No matter their age, students should play an important ownership role in accessing, organizing, and caring for lab supplies and equipment.

Internet-connected devices are invaluable to many aspects of science learning. Many schools are one-to-one technology schools in which every student has a school-issued tablet or laptop device. In schools that don't integrate such programs, laptop carts or computer labs may be available. If all students have access to smartphones, they can generally use these devices for many investigations. For example, most Vernier applications (software associated with probeware that can measure things like light wavelengths, pH, and dissolved oxygen) can now be run on smartphones and connect to data-logging devices. In addition, phones, tablets, and laptops can all run the Google software suite noted in many elements throughout this book.

Placing Student Desks and Planning Areas for Group Work

Ideally, a science classroom will have furniture that supports and encourages cooperative learning. In a classroom of twenty-four to twenty-eight students, arrange six to seven tables or pods that seat four students each to allow proper spacing based on the room's footprint. In addition, orient tables so every student has a clear view to a projector screen or the most common locations where the teacher presents information and lessons. If possible, both table legs and chair legs should have casters or other feet to allow easy movement to create flexible groupings. If tabletops are black and sealed, they also make great surfaces for writing and drawing with neon dry-erase markers.

Some schools only have individual student desks with attached seats. Tennis balls may be added to desk legs so they can slide easily across the floor. The desks may be grouped in fours to facilitate conversations between face partners and shoulder partners. Regardless of the current state of the room, teachers should assess the footprint and the versatility of the furniture. Following the assessment, teachers should work with housekeeping and janitorial colleagues to maximize the flexibility of the learning space.

Element 39: Understanding Students' Backgrounds and Interests

Understanding students' backgrounds and interests goes a long way in developing positive relationships between teachers and students. When teachers conduct activities that uncover each student's accomplishments, likes, and dislikes, students feel more valued as part of the group. In a 2021 study, the perception of being valued by the teacher and motivation for learning were positively correlated (Bostan, Stanciu, & Andronic, 2021). Student backgrounds and interests also are useful in the learning process for exploring scientific phenomena. In scientific discovery processes, it is useful for students to make predictions. They typically cannot begin to do this without being able to draw from the reservoir of their life experiences. If teachers are able to learn about their students and incorporate who they are and what they might know into the curriculum, the learning experience becomes much richer.

This section presents strategies for understanding students' backgrounds and interests. Figure 8.2 depicts the self-rating scale for this element so teachers can gauge their performance.

Score	Description
4: Innovating	I engage in all behaviors at the Applying level. In addition, I identify those students who do not appear to perceive that the teacher genuinely likes them and design alternate activities and strategies to meet their specific needs.
3: Applying	I engage in activities to understand students and monitor the extent to which students perceive that I am genuinely interested in them.
2: Developing	I engage in verbal and nonverbal behaviors to indicate affection for students without significant errors or omissions.
1: Beginning	I engage in activities to understand students' backgrounds and interests but do so with errors or omissions, such as exhibiting these behaviors with some but not all students and engaging in these behaviors in a perfunctory manner.
0: Not Using	I do not engage in activities to understand students' backgrounds and interests.

Figure 8.2: Self-rating scale for element 39—Understanding students' backgrounds and interests.

Of the assortment of strategies for understanding students' backgrounds and interests that we identify in appendix A (page 149), we elaborate on the following specific strategies in this section.

- Student background surveys
- Opinion questionnaires
- Familiarity with student culture
- Six-word autobiographies

Student Background Surveys

To get to know students personally, many teachers decide to administer a survey at the beginning of the year to ascertain cultural backgrounds, hobbies, plans for the future, and a variety of other pieces of information valuable to forming a solid teacher-student relationship. The information gathered is only helpful if the teacher thoroughly reads student responses and reflects back to students the information in a way that they know they are being seen and heard as unique individuals.

At the beginning of the year, author Brett Erdmann gives a survey to his students via a Google form, as shown in figure 8.3.

Last name	
First name (what I like to be called in class)	
Personal pronouns	
School ID#	
Birthday	
Parent or guardian email	
Clubs, activities, sports	
Three of my favorite things	

Figure 8.3: Student survey.

continued →

One thing I want my teacher to know about me	
I enjoy learning when . . .	
What's your sentence? Please answer this simple question in a short video (one to two minutes). Be sure to introduce yourself by name. Tell me your sentence and then spend a little bit elaborating on your core values and how they relate to your sentence. Paste a shared (public) link from YouTube here. I will view this, and it will not be shared in class.	

Visit **go.SolutionTree.com/instruction** for a free reproducible version of this figure.

When collecting surveys, it is critical to communicate what you will do with the information. If teachers are collecting information only to help improve the teacher-student relationship, they should state this and agree that the information is not public. Teachers should also be honest and stress that they are mandated public reporters and might have to share anything that is a potential safety issue. If a teacher collects information that they may wish to share with a parent, then those parts of the survey should also be clearly communicated.

In addition to collecting personal information, surveys can help with instructional planning. For example, if a teacher knows that many students watch a certain television show, he or she might build in connections between a lesson and the show's topic, such as forensic science. A teacher may also learn that several students create video content on YouTube and TikTok, which could inform the way that teacher structures some activities to include more student video productions in conjunction with the explanation of scientific claims. Specifically, some students may be high volume or widely followed content creators on these channels. These creators can teach other students (and teachers) many useful tools and strategies to make these videos compelling and visually appealing.

Opinion Questionnaires

Opinion questionnaires target student interest in specific classroom science content or investigational approaches. These questionnaires allow teachers to collect information to better understand what students would like to know and do in the context of the curriculum.

If teachers know that students have specific science content interests at the beginning of the year, they can plan to incorporate those into their lessons or units. For example, a science teacher may find that several students have a deep interest in the human impact on climate change. This information may inform the teacher's approach to the teaching of units related to photosynthesis, cellular respiration, ecology, or energy transfer.

An opinion questionnaire might include the following types of questions.

- On a scale of 1 to 5, rate your comfort level with identifying and representing the various elements of good experimental design.
- What are your favorite science (or biology, chemistry, and so on) topics?

- What is one question you want this class to answer by the end of the year?
- What are you really looking forward to in this class?
- What are your fears as you enter this class?
- Do you prefer to receive feedback from the teacher orally, in written form on the assignment, by email, or by other means?
- In which of the following areas are you strongest: computational thinking, conceptual modeling, argumentation, or explanation (claim, evidence, reasoning)?
- In which of the following areas are you weakest: computational thinking, conceptual modeling, argumentation, or explanation (claim, evidence, reasoning)?
- If you had to choose one scientific or STEM field as a career, what would it be? Why?

Familiarity With Student Culture

When using culturally responsive teaching practices, educators find ways to have students share about their neighborhoods, cultures, traditions, and family histories. This goes well beyond any information you might collect in a student background survey. To be truly culturally responsive, students must have frequent opportunities to share their lived experiences within the context of regular learning cycles. Sharing becomes embedded and exposes the varying frames of reference we use to view the world and the special skills (and biases) we bring to the exploration of scientific phenomena.

Through this sharing, teachers and students become cognizant that everyone comes to class with slightly different lenses through which they view the world. Once the teacher identifies these lenses, he or she can apply them to classroom learning experiences. Science teaching has long been Eurocentric, focusing on the work of a small number of white males. The story of science is populated by people of color, women, and a great diversity of nations and cultures. When teachers only focus on a narrow narrative, students who aren't represented have trouble seeing themselves as meaningful contributors to STEM fields.

To be inclusive, teachers should focus on both the simple and complex. They can write assessments and word problems that include names from various cultures and aren't historically Caucasian, such as Suzie and Billy. Instead, a teacher may wish to use Laquita or Lonzo, thus using names that would recognize the cultures present in the class. Teachers should also be cognizant of using metaphors and examples that appeal to females and students of color.

When highlighting the history of science, teachers will certainly include Isaac Newton, Albert Einstein, and Charles Darwin. However, there are great stories to share about women and people of color, such as Barbara McClintock, Hedy Lamarr, Katherine Johnson, Percy Julian, and Marie Curie (to name a few). Discussions also should include international scientists such as Imran Ali, Shinya Yamanka, Seiji Ogawa, and C. Nüsslein-Volhard.

More complicated integrations of culture may involve classroom discussions of the connections (and lack thereof) between race and biology (mentioned in an earlier learning strategy about Venn diagrams). One very compelling discussion in science classes revolves around the concept that race is a cultural construct and the biological differences between racial groups are minimal and relegated mainly to particular disease propensities or sensitivities. The discrimination that minority groups face is rooted in culture and not biology. Although scientific discovery is often viewed as purely objective, it is often quite the opposite and represents only the product of a certain time, place, and group of people.

Six-Word Autobiographies

Six-word autobiographies are very much like an elevator pitch applied to a student's personal history. These six-word autobiographies allow students to share their backgrounds in a way that can begin the deeper sharing that happens in the science learning. This simple exercise allows students to try to encapsulate their lives in no more than six words. The six words may be a complete sentence or a list of adjectives describing their personal history. Teachers can display students' six-word autobiographies publicly on the wall in conjunction with an earlier described strategy in this chapter (displaying student work). These autobiographies also can be shared privately with the teacher in digital format and, in this manner, can be continuously revisited and revised throughout the course of the school year.

In his book *Drive*, Daniel Pink (2009) describes a related version of this strategy called What's Your Sentence? Basically, this is a thinking exercise in which students write about themselves in the past tense. They answer the question, "What would you want people to say about you after you die?" In essence, the answer is a personal mission statement. A student might say, "She always worked with dancers to help them achieve their highest potential." Another student might say, "He loved animals and invented special foods so they could live healthy and happy lives." This strategy is a little different than the simple six-word autobiography because it makes students consider their future as part of the autobiographical storytelling.

GUIDING QUESTIONS FOR CURRICULUM DESIGN

The design questions in this chapter focus on implementing rules and procedures and building relationships: *What strategies will I use to help students understand and follow rules and procedures?* and *How will I demonstrate that I understand students' backgrounds and interests?* The following questions, which align to each of the elements in this chapter, guide teachers to organize learning spaces effectively and build relationships.

- **Element 34:** What will I do to make the physical layout of the classroom most conducive to learning?

_____

_____

_____

_____

_____

_____

_____

_____

_____

_____

_____

_____

_____

_____

- **Element 39:** What strategies will I use to help students feel welcome, accepted, and valued?

_____

_____

_____

_____

_____

_____

_____

_____

_____

_____

_____

_____

_____

Summary

Strategies associated with the physical layout of the science classroom allow easy movement around the classroom, make materials accessible, showcase students' work, and account for technology equipment. Being conscious of the physical layout of the room also promotes and ensures safe science investigations. By attending to this physicality, teachers can enhance students' sense of order, which contributes to their learning. Additionally, teachers can seek to understand students' backgrounds and interests, which help build students' perception that their teacher and peers respect and value them. This, too, can elevate learning as students feel valued, connected, and see themselves in the scientists discussed and recognized in class. The next chapter offers suggestions to develop teachers' competency and expertise in the art and science of this profession.

CHAPTER 9

Developing Expertise

As the previous chapters illustrate, *The New Art and Science of Teaching* is a comprehensive framework that can help science teachers develop their expertise and, in turn, increase student learning. This relationship between teacher expertise and student learning is well established. For example, in one of the most rigorous studies of the relationship between teacher effectiveness and student achievement, Barbara Nye, Spyros Konstantopoulos, and Larry V. Hedges (2004) estimate that the difference in student achievement between a teacher who is "average" and a teacher who is "very effective" is about 13 percentile points in reading and 18 in mathematics. They note, "These effects are certainly large enough effects to have policy significance" (Nye et al., 2004, p. 253).

In *Visible Learning*, Hattie (2009) dedicates a chapter to the impact teachers have on student learning gains. In particular, his research highlights the larger effect sizes related to teacher expectations, teacher professional development, teacher clarity, and quality of teaching as having significant impacts on student learning and growth. In more recent work, Hattie (2017) finds that collective teacher efficacy, or the shared belief among teachers that their collective efforts can and will improve student learning, is the single largest factor in improving student learning gains.

This relationship between teacher effectiveness and student learning underscores the importance of helping teachers continually develop their expertise. The more skilled teachers become, the more their students learn. It's as simple as that. Fortunately, there is a great deal of research and theory on how to improve one's expertise in any complex domain such as teaching and assessment (Cooper, 2022; Ericsson & Charness, 1994; Ericsson, Krampe, & Tesch-Römer, 1993; Lang, 2021; Marzano, 2010; Marzano & Eaker, 2020).

One of the most powerful aspects of *The New Art and Science of Teaching* framework is its granularity. Simply put, the framework describes effective teaching in such detail that educators can pinpoint areas of strength and weakness, set goals to improve in specific areas, select strategies to implement as they seek to improve, track progress as they go, and make subsequent plans for future growth. In this book, we have specifically applied the forty-three elements of *The New Art and Science of Teaching* framework to science classrooms. Thus, science teachers can use the guidance from chapters 1 through 8, along with the reflective process detailed in this chapter, to enhance their professional practice.

The purpose of this reflective process is for teachers to regularly and routinely assess strengths and opportunities for growth in their professional practice, particularly when compared to the elements and strategies explained in this book. With this reflection in place, science teachers can then formulate specific and manageable steps to learn and bring new strategies into their practice.

The reflective process commences with a *self-audit*. Using a developmental scale such as the one shown in figure 9.1 (and found throughout this book as a part of each element), teachers rate themselves on the forty-three elements of *The New Art and Science of Teaching* framework.

1. Conduct a self-audit.
2. Select goal elements and specific strategies.
3. Engage in deliberate practice and track progress.
4. Seek continuous improvement by planning for future growth.

Step 1: Conduct a Self-Audit

The first step in the reflective process is to conduct a *self-audit*. The purpose of conducting a self-audit is to determine one's strengths and weaknesses. Strengths are celebrated; weaknesses become the focus of developing expertise. Science teachers can easily accomplish this using the self-rating scales that we present for each element covered in this text. Recall that we designed each of those scales in the format shown in figure 9.1.

Score	Description
4: Innovating	I adapt behaviors and create new strategies for unique student needs and situations.
3: Applying	I use the strategies and behaviors associated with this element, and I monitor the extent to which my actions affect students' performance.
2: Developing	I use the strategies and behaviors associated with this element, but I do not monitor the effect on students.
1: Beginning	I use the strategies and behaviors associated with this element incorrectly or with parts missing.
0: Not Using	I am unaware of strategies and behaviors associated with this element.

Source: Marzano, 2017, p. 104.

Figure 9.1: Developmental scale for elements.

As noted in the introduction (page 1), customized scales accompany the elements in this book. We recommend that each year, teachers score themselves using these scales. In effect, teachers should construct a profile of their strengths and weaknesses. They can consider any elements on which they have scores of Applying or Innovating as strengths. Any elements on which teachers have scores of Not Using, Beginning, or Developing are candidates for improvement. Visit **go.SolutionTree.com/instruction** to download free reproducible versions of all the scales.

Teachers can use the self-audit form in figure 9.2 to record the results from the self-audit. Note that while not all elements in the figure are specifically addressed in this book, they have been included for reference and reflection.

Element	4	3	2	1	O
1. Providing scales and rubrics					
2. Tracking student progress					
3. Celebrating success					
4. Using informal assessments of the whole class					
5. Using formal assessments of individual students					
6. Chunking content					
7. Processing content					
8. Recording and representing content					
9. Using structured practice sessions					
10. Examining similarities and differences					
11. Examining errors in reasoning					
12. Engaging students in cognitively complex tasks					
13. Providing resources and guidance					
14. Generating and defending claims					
15. Previewing					
16. Highlighting critical information					
17. Reviewing content					
18. Revising knowledge					
19. Reflecting on learning					
20. Assigning purposeful homework					
21. Elaborating on information					
22. Organizing students to interact					
23. Noticing and reacting when students are not engaged					
24. Increasing response rates					
25. Using physical movement					
26. Maintaining a lively pace					
27. Demonstrating intensity and enthusiasm					
28. Presenting unusual information					
29. Using friendly controversy					
30. Using academic games					
31. Providing opportunities for students to talk about themselves					
32. Motivating and inspiring students					
33. Establishing rules and procedures					
34. Organizing the physical layout of the classroom					
35. Demonstrating withitness					
36. Acknowledging adherence to rules and procedures					
37. Acknowledging lack of adherence to rules and procedures					
38. Using verbal and nonverbal behaviors that indicate affection for students					
39. Understanding students' backgrounds and interests					
40. Displaying objectivity and control					
41. Demonstrating value and respect for reluctant learners					
42. Asking in-depth questions of reluctant learners					
43. Probing incorrect answers with reluctant learners					

Figure 9.2: Self-audit for *The New Art and Science of Teaching* framework.

*Visit **go.SolutionTree.com/instruction** for a free reproducible version of this figure.*

Step 2: Select Goal Elements and Specific Strategies

Once teachers have conducted a self-audit, they identify elements they will focus on for their personal and professional development over the upcoming year. We recommend that teachers choose from one to three elements per year. If a supervisor, evaluator, or instructional coach is involved in the process, this individual should have some say in what a teacher selects. That is, he or she might ask a beginning or struggling teacher to work on two elements and allow the teacher to select one. Sometimes the school or district has adopted a focus, in which case one element might be a collective mandate, and teachers choose the others pertinent to their level of expertise and areas needing improvement. To illustrate, assume a teacher selects the following elements on which she has scored herself as Developing (2), Beginning (1), and Not Using (0) respectively.

- **Element 6:** Chunking content
- **Element 11:** Examining errors in reasoning
- **Element 16:** Highlighting critical information

While there might be a number of other elements for which the teacher has assigned herself a relatively low score, these are the ones she wishes to focus on in the coming year. At a more granular level, the teacher then identifies specific strategies to target as a means of improvement. Recall that each element contains multiple strategies; for example, element 16 (highlighting critical information) encompasses the strategies listed below. This book specifically addressed the following strategies in this element: repeating the most important content, using visual activities, and using dramatic instructions to convey critical content. The others are listed for reference.

- Repeating the most important content
- Asking questions that focus on critical information
- Using visual activities
- Using narrative activities
- Using tone of voice, gestures, and body position
- Using pause time
- Identifying critical-input experiences
- Using explicit instruction to convey critical content
- Using dramatic instruction to convey critical content
- Providing advance organizers to cue critical content
- Using what students already know to cue critical content

From this list, the teacher might select the following strategies: (1) repeating the most important content and (2) using pause time.

Step 3: Engage in Deliberate Practice and Track Progress

After identifying specific strategies within each selected focus element, the teacher engages in deliberate practice. As its name implies, *deliberate practice* involves focusing on explicit goals and monitoring incremental progress toward those goals While there are some gains through deliberate practice, and it is helpful in acquiring skills, one of the real benefits is for the teacher to see the various strategies available and begin incorporating them into the classroom (Hattie, 2017; Macnamara, Hambrick, & Oswald, 2014). Again, the self-rating scales for the chosen elements should be of service when gauging progress.

In our example, the teacher initially scored herself as Not Using the element of highlighting critical information. She then handpicked the strategy: repeating the most important content. This implies that she really doesn't even know how the strategy might manifest itself in the classroom. Therefore, she incorporates the strategies listed in this book, engages in some internet searches to find a few examples described in downloadable documents, and identifies some video examples of how teachers use this strategy. Additionally, she might ask colleagues how they have incorporated this strategy in their classrooms with success. Once she decides that she has enough information about the strategy, she is ready to try it out in her own classroom. She now has moved to the Beginning level on the self-rating scale.

The Beginning level means that a teacher is attempting to use a strategy but makes some significant errors. On a day the teacher tries out the strategy, she might solicit the support of a colleague—such as an instructional coach, fellow teacher, or administrator—to observe her with the express purpose of offering feedback on the use of this target strategy. She implements the strategy again and puts aside time to reflect on what she did well and what she could have done better. After about four attempts at using the strategy, along with the self-rating scale, the teacher concludes that she is using the strategy without making any significant errors. She is now at the Developing level.

To move to the Applying level, the teacher asks herself, "What do I expect to see students doing if this strategy is working well?" The answer is obvious. After she presents new information to students, they should understand and remember what is most important out of all the information she presents to them. To obtain a sense of this, she decides to ask students to respond to exit slips with the question, *What was most important about what we covered in today's class?* To her pleasant surprise, most of the students can accurately identify the critical content. She is now at the Applying level.

To advance to the Innovating level, the teacher examines the reactions of her students in even more depth by briefly talking to students who still seem to have trouble identifying critical information, even after she uses the strategy. She soon realizes that many of these students are English learners, and simply repeating the content doesn't help them get over the language barrier. She decides to include some graphic and pictographic representations of the content, along with her verbal repetitions, and soon finds that it helps bridge the gap for these students.

A tracking chart like that in figure 9.3 can help teachers track their progress through each level of the scale.

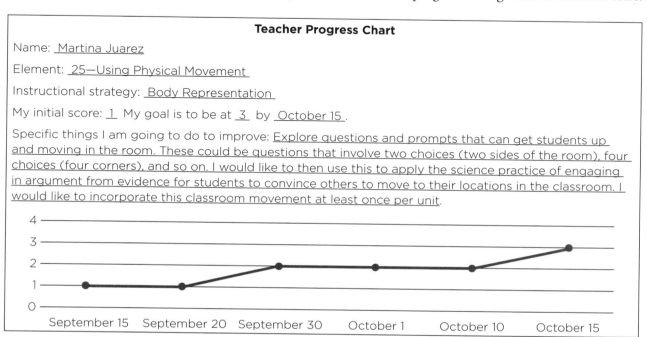

Figure 9.3: Teacher progress for the strategy of using physical movement—Body representation.

Step 4: Engage in Continuous Improvement by Planning for Future Growth

The final step in the development process is for teachers to keep selecting elements and strategies on which to work and improving their practice in those elements and strategies. Even if a teacher becomes competent in all elements of *The New Art and Science of Teaching* model, there are still multiple strategies within each element on which to improve. In effect, the model can provide teachers with new challenges and new levels of expertise throughout their entire career.

Summary

Developing expertise is not a function of talent or serendipity. Rather, it is a product of focus and hard work over time. The four steps we provide in this chapter should address the criterion of focus. The criterion of hard work over time is a product of teacher commitment. Hopefully, the resources provided in this book help elicit such commitment.

Afterword

The New Art and Science of Teaching presents a comprehensive model of teaching that organizes all or most of the instructional strategies available to teachers. The science reference is predicated on the fact that these strategies are founded on decades of research and theory and contribute to effective teaching. The art component indicates that factors other than research attribute to student learning, such as which strategies are used together and how teachers use them for express purposes. This analogy can help elucidate this point:

> Instructional strategies are best likened to techniques an artist might develop and refine over years of practice. The artist then uses these techniques to create works that are not only unique and complex but elegantly focused. The more skill the artist exhibits with available techniques, the better his or her creations. Likewise, the more skill the classroom teacher has with the instructional strategies that research and theory have uncovered over the decades, the better the teacher will be able to create lessons that optimize student learning. (Marzano, 2017, p. 2)

It is the duty, the call to action, and the mission of educators everywhere to meet students where they are and elevate them to the next level of learning. This is an awesome task indeed and extremely rewarding when students move from not knowing to awareness.

As teachers endeavor to undertake this responsibility, they need tools, resources, and support to help guide them so they can be the best possible conduit of learning for their charges. *The New Art and Science of Teaching Science* presents myriad strategies to assist teachers in this work. We invite all teachers to raise their own bar of professional capacity so they can, in turn, open the door for the students they are so fortunate to lead.

Appendix A
The New Art and Science of Teaching Framework Overview

As explained in the introduction (page 1), *The New Art and Science of Teaching* framework involves three overarching categories—(1) feedback, (2) content, and (3) context. We have further divided each of these categories into ten design areas, each of which corresponds with a specific teacher action. The three categories give rise to forty-three elements which, on a more granular level, comprise 333 associated instructional strategies. Figure A.1 (page 150) presents a comprehensive list of the forty-three elements and their associated strategies as they appear in *The New Art and Science of Teaching* (Marzano, 2017). The teacher actions, elements, and instructional strategies that appear in bold typeface are those that we feature in this book as they relate to science instruction.

The New Art and Science of Teaching Framework Overview

Category	Teacher Actions	Desired Student Mental States and Processes	Design Questions	Forty-Three Elements (We address the bolded elements in this book.)	Strategies (We address the bolded strategies in this book.)
Feedback	Chapter 1: Providing and Communicating Clear Learning Goals	Students understand the progression of knowledge teachers expect them to master and where they are along that progression.	How will I communicate clear learning goals that help students understand the progression of knowledge I expect them to master and where they are along that progression?	1. Providing scales and rubrics How will I design scales or rubrics?	1. Clearly articulating learning goals 2. Creating scales or rubrics for learning goals 3. Implementing routines for using targets and scales 4. Using teacher-created targets and scales 5. Creating student-friendly scales 6. Identifying individual student learning goals
				2. Tracking student progress How will I track progress?	7. Using formative scores 8. Designing assessments that generate formative scores 9. Using individual score-level assessments 10. Using different types of assessments 11. Generating summative scores 12. Charting student progress 13. Charting class progress
				3. Celebrating success How will I celebrate success?	14. Status celebration 15. Knowledge gain celebration 16. Verbal feedback
	Chapter 2: Using Assessments	Students understand how test scores and grades relate to their status on the progression of knowledge teachers expect them to master.	How will I design and administer assessments that help students understand how their test scores and grades relate to their status on the progression of knowledge I expect them to master?	4. Using informal assessments of the whole class How will I informally assess the whole class?	17. Confidence rating techniques 18. Voting techniques 19. Response boards 20. Unrecorded assessments
				5. Using formal assessments of individual students How will I formally assess individual students?	21. Common assessments designed using proficiency scales 22. Assessments involving selected-response or short constructed-response items 23. Student demonstrations 24. Student interviews 25. Observations of students 26. Student-generated assessments 27. Response patterns

Content				
Chapter 3: Conducting Direct Instruction Lessons	When content is new, students understand which parts are important and how the parts fit together.	When content is new, how will I design and deliver direct instruction lessons that help students understand which parts are important and how the parts fit together?	**6. Chunking content** How will I chunk the new content into short, digestible bites?	28. **Using preassessment data to plan for chunks** 29. **Presenting content in small, sequentially-related sets** 30. **Allowing for processing time between chunks**
			7. Processing content How will I help students process the individual chunks and the content as a whole?	31. **Perspective analysis** 32. **Thinking hats** 33. Collaborative process 34. Jigsaw cooperative learning 35. Reciprocal teaching 36. **Concept attainment** 37. Think-pair-share 38. Scripted cooperative dyads
			8. Recording and representing content How will I help students record and represent their knowledge?	39. Informal outlines 40. **Summaries** 41. Pictorial notes and pictographs 42. Combination notes, pictures, and summaries 43. Graphic organizers 44. **Free-flowing webs** 45. Academic notebooks 46. **Dramatic enactments** 47. **Mnemonic devices** 48. Rhyming peg words 49. Link strategies
Chapter 4: Conducting Practicing and Deepening Lessons	After teachers present new content, students deepen their understanding and develop fluency in skills and processes.	After I have presented content, how will I design and deliver lessons that help students deepen their understanding and develop fluency in skills and processes?	**9. Using structured practice sessions** How will I use structured practice?	50. **Modeling** 51. **Guided practice** 52. Close monitoring 53. Frequent structured practice 54. **Varied practice** 55. Fluency practice 56. Worked examples 57. Practice sessions prior to testing

Source: Adapted from Glass & Marzano, 2018.

Figure A.1: *The New Art and Science of Teaching* framework overview.

continued →

The New Art and Science of Teaching Framework Overview

Category	Teacher Actions	Desired Student Mental States and Processes	Design Questions	Forty-Three Elements	Strategies
			After I have presented content, how will I design and deliver lessons that help students deepen their understanding and develop fluency in skills and processes?	**10. Examining similarities and differences** How will I help students examine similarities and differences?	58. Sentence-stem comparisons 59. Summaries **60. Constructed-response comparisons** **61. Venn diagrams** 62. T-charts 63. Double-bubble diagrams 64. Comparison matrices 65. Classification charts **66. Dichotomous keys** **67. Sorting, matching, and categorizing** 68. Similes 69. Metaphors 70. Sentence-stem analogies 71. Visual analogies
				11. Examining errors in reasoning How will I help students examine errors in reasoning?	72. Identifying errors of faulty logic 73. Identifying errors of attack 74. Identifying errors of weak reference **75. Identifying errors of misinformation** 76. Practicing identifying errors in logic **77. Finding errors in the media** 78. Examining support for claims 79. Judging reasons and evidence in an author's work 80. Identifying statistical limitations 81. Using student-friendly prompts 82. Anticipating student errors 83. Avoiding unproductive habits of mind
Chapter 5: Conducting Knowledge Application Lessons		After teachers present new content, students generate and defend claims through knowledge application tasks.	After I have presented content, how will I design and deliver lessons that help students generate and defend claims through knowledge application?	**12. Engaging students in cognitively complex tasks** How will I engage students in cognitively complex tasks?	**84. Experimental-inquiry tasks** 85. Problem-solving tasks 86. Tasks to examine the efficiencies of multiple methods of problem solving 87. Decision-making tasks 88. Investigation tasks **89. Invention tasks** 90. Student-designed tasks

Chapter		Element	Strategies
	After I have presented content, how will I design and deliver lessons that help students generate and defend claims through knowledge application?	**13. Providing resources and guidance** How will I provide resources and guidance?	91. Using proficiency or scoring scales **92. Providing resources** 93. Providing informational handouts 94. Teaching research skills 95. Conducting interviews **96. Circulating around the room** 97. Collecting informal assessment information **98. Offering feedback** **99. Creating cognitive dissonance**
		14. Generating and defending claims How will I help students generate and defend claims?	100. Introducing the concept of claims and support 101. Presenting the formal structure of claims and support 102. Generating claims **103. Providing grounds** 104. Providing backing 105. Generating qualifiers **106. Formally presenting claims**
Chapter 6: Using Strategies That Appear in All Types of Lessons	Students continually integrate new knowledge with old knowledge and revise their understanding accordingly.	**15. Previewing strategies** How will I help students preview content? Throughout all types of lessons, what strategies will I use to help students continually integrate new knowledge with old knowledge and revise their understanding accordingly?	**107. Informational hooks** 108. Bell ringers **109. What do you think you know?** 110. Overt linkages **111. Preview questions** 112. Brief teacher summaries 113. Skimming 114. Teacher-prepared notes 115. K-W-L strategies 116. Advance organizers 117. Anticipation guides 118. Word splashes 119. Preassessments

continued →

The New Art and Science of Teaching Framework Overview

Category	Teacher Actions	Desired Student Mental States and Processes	Design Questions	Forty-Three Elements	Strategies
			Throughout all types of lessons, what strategies will I use to help students continually integrate new knowledge with old knowledge and revise their understanding accordingly?	**16. Highlighting critical information** How will I highlight critical information?	**120. Repeating the most important content** 121. Asking questions that focus on critical information **122. Using visual activities** 123. Using narrative activities 124. Using tone of voice, gestures, and body position 125. Using pause time 126. Identifying critical input experiences 127. Using explicit instruction to convey critical content **128. Using dramatic instruction to convey critical content** 129. Providing advance organizers to cue critical content 130. Using what students already know to cue critical content
				17. Reviewing content How will I help students review content?	**131. Cumulative review** 132. Cloze activities 133. Summary 134. Presented problem **135. Demonstration** 136. Brief practice test or exercise 137. Questioning **138. Give one, get one**
				18. Revising knowledge How will I help students revise knowledge?	139. Academic notebook entries 140. Academic notebook review **141. Peer feedback** **142. Assignment revision** 143. The five basic processes 144. Visual symbols **145. Writing tools**
				19. Reflecting on learning How will I help students reflect on their learning?	146. Reflective journals 147. Think logs **148. Exit slips** **149. Knowledge comparisons** 150. Two-column notes

20. Assigning purposeful homework How will I use purposeful homework?	Throughout all types of lessons, what strategies will I use to help students continually integrate new knowledge with old knowledge and revise their understanding accordingly?	151. Homework preview 152. Homework to deepen knowledge **153. Homework to practice a process or skill** 154. Parent-assessed homework
21. Elaborating on information How will I help students elaborate on information?		155. General inferential questions **156. Elaborative interrogation** 157. Questioning sequences
22. Organizing students to interact How will I organize students to interact?		158. Group for active processing **159. Group norms creation** 160. Fishbowl demonstration 161. Job cards 162. Predetermined buddies to help form ad hoc groups 163. Contingency plan for ungrouped students **164. Group using preassessment information** 165. Pair-check **166. Think-pair-share and think-pair-square** 167. Student tournaments 168. Inside-outside circle 169. Cooperative learning **170. Peer-response groups** 171. Peer tutoring 172. Structured grouping **173. Group reflection on learning**

continued →

The New Art and Science of Teaching Framework Overview

Category	Teacher Actions	Desired Student Mental States and Processes	Design Questions	Forty-Three Elements	Strategies
Context	Chapter 7: Using Engagement Strategies	Students are paying attention, energized, intrigued, and inspired.	What engagement strategies will I use to help students pay attention, be energized, be intrigued, and be inspired?	23. Noticing and reacting when students are not engaged What will I do to notice when students are not engaged, and how will I react?	174. Monitoring individual student engagement 175. Monitoring overall class engagement 176. Using self-reported student engagement data 177. Re-engaging individual students 178. Boosting overall class energy levels
				24. Increasing response rates What will I do to increase students' response rates?	179. Random names 180. Hand signals 181. Response cards 182. Response chaining 183. Paired response 184. Choral response 185. Wait time 186. Elaborative interrogation 187. Multiple types of questions
				25. Using physical movement What will I do to increase students' physical movements?	188. Stand up and stretch 189. Vote with your feet 190. Corners activities 191. Stand and be counted 192. Body representations 193. Drama-related activities
				26. Maintaining a lively pace What will I do to maintain a lively pace?	194. Instructional segments 195. Pace modulation 196. The parking lot 197. Motivational hooks

continued →

What engagement strategies will I use to help students pay attention, be energized, be intrigued, and be inspired?

27. Demonstrating intensity and enthusiasm What will I do to demonstrate intensity and enthusiasm?	**28. Presenting unusual information** What will I do to present unusual information?	**29. Using friendly controversy** What will I do to engage students in friendly controversy?	**30. Using academic games** How will I use academic games to engage students?
198. Direct statements about the importance of content 199. Explicit connections **200. Nonlinguistic representations** **201. Personal stories** 202. Verbal and nonverbal signals 203. Humor 204. Quotations 205. Movie and film clips	206. Teacher-presented information 207. WebQuests **208. Fast facts** 209. Believe it or not **210. History files** **211. Guest speakers**	212. Friendly controversy 213. Class vote **214. Seminars** 215. Expert opinions 216. Opposite point of view **217. Diagrams comparing perspectives** 218. Lincoln–Douglas debate 219. Town-hall meeting 220. Legal model	221. What is the question 222. Name that category 223. Talk a mile a minute 224. Classroom feud **225. Which one doesn't belong?** **226. Inconsequential competition** 227. Turning questions into games **228. Vocabulary review games**

The New Art and Science of Teaching Framework Overview

Category	Teacher Actions	Desired Student Mental States and Processes	Design Questions	Forty-Three Elements	Strategies
Context			What engagement strategies will I use to help students pay attention, be energized, be intrigued, and be inspired?	**31. Providing opportunities for students to talk about themselves** How will I provide opportunities for students to talk about themselves?	**229. Interest surveys** **230. Student learning profiles** 231. Life connections 232. Informal linkages during class discussion
				32. Motivating and inspiring students How will I motivate and inspire students?	233. Academic goal setting 234. Growth mindset cultivation **235. Possible selves activities** 236. Personal projects 237. Altruism projects **238. Gratitude journals** 239. Mindfulness practice 240. Inspirational media
	Chapter 8: Implementing Rules and Procedures	Students understand and follow rules and procedures.	What strategies will I use to help students understand and follow rules and procedures?	33. Establishing rules and procedures What will I do to establish rules and procedures?	241. Using a small set of rules and procedures 242. Explaining rules and procedures to students 243. Generating rules and procedures with students 244. Modifying rules and procedures with students 245. Reviewing rules and procedures with students 246. Using the language of responsibility and statements of school beliefs 247. Posting rules around the room 248. Writing a class pledge or classroom constitution 249. Using posters and graphics 250. Establishing gestures and symbols 251. Modeling with vignettes and role playing 252. Holding classroom meetings 253. Implementing student self-assessment

Strategy / Question	Items
What strategies will I use to help students understand and follow rules and procedures?	
34. Organizing the physical layout of the classroom What will I do to make the physical layout of the classroom most conducive to learning?	254. Designing classroom décor **255. Displaying student work** **256. Considering classroom materials** 257. Placing the teacher's desk **258. Placing student desks** 259. Planning areas for whole-group instruction **260. Planning areas for group work** 261. Planning learning centers **262. Considering computers and technology equipment** 263. Considering lab equipment and supplies 264. Planning classroom libraries 265. Involving students in the design process
35. Demonstrating withitness What will I do to demonstrate withitness?	266. Being proactive 267. Occupying the whole room physically and visually 268. Noticing potential problems 269. Using a series of graduated actions
36. Acknowledging adherence to rules and procedures What will I do to acknowledge adherence to rules and procedures?	270. Verbal affirmation 271. Nonverbal affirmation 272. Tangible recognition 273. Token economies 274. Daily recognition form 275. Color-coded behavior 276. Certificates 277. Phone calls, emails, and notes
37. Acknowledging lack of adherence to rules and procedures What will I do to acknowledge lack of adherence to rules and procedures?	278. Verbal cues 279. Pregnant pause 280. Nonverbal cues 281. Time-out 282. Overcorrection 283. Interdependent group contingency 284. Home contingency 285. High-intensity situation plan 286. Overall disciplinary plan

continued →

The New Art and Science of Teaching Framework Overview

Category	Teacher Actions	Desired Student Mental States and Processes	Design Questions	Forty-Three Elements	Strategies
Context	**Chapter 9: Building Relationships**	Students feel welcome, accepted, and valued.	What strategies will I use to help students feel welcome, accepted, and valued?	38. Using verbal and nonverbal behaviors that indicate affection for students How will I use verbal and nonverbal behaviors that indicate affection for students?	287. Greeting students at the classroom door 288. Holding informal conferences 289. Attending after-school functions 290. Greeting students by name outside of school 291. Giving students special responsibilities or leadership roles in the classroom 292. Scheduling interaction 293. Creating a photo bulletin board 294. Using physical behaviors 295. Using humor
				39. **Understanding students' backgrounds and interests** How will I demonstrate that I understand students' backgrounds and interests?	296. **Student background surveys** 297. **Opinion questionnaires** 298. Individual student-teacher conferences 299. Parent-teacher conferences 300. School newspaper, newsletter, or bulletin 301. Informal class interviews 302. **Familiarity with student culture** 303. Autobiographical metaphors and analogies 304. **Six-word autobiographies** 305. Independent investigations 306. Quotes 307. Comments about student achievement or areas of importance 308. Lineups 309. Individual student learning goals
				40. Displaying objectivity and control How will I demonstrate objectivity and control?	310. Self-reflection 311. Self-monitoring 312. Emotional triggers 313. Self-care 314. Assertiveness 315. A cool exterior 316. Active listening and speaking 317. Communication styles 318. Unique student needs

Chapter 10: Communicating High Expectations	Typically reluctant students feel valued and do not hesitate to interact with the teacher or their peers.	What strategies will I use to help typically reluctant students feel valued and comfortable interacting with me or their peers?		
		41. Demonstrating value and respect for reluctant learners How will I demonstrate value and respect for reluctant learners?	319. Identifying expectation levels for all students 320. Identifying differential treatment of reluctant learners 321. Using nonverbal and verbal indicators of respect	
		42. Asking in-depth questions of reluctant learners How will I ask in-depth questions of reluctant learners?	322. Question levels 323. Response opportunities 324. Follow-up questioning 325. Evidence and support for student answers 326. Encouragement 327. Wait time 328. Tracking responses 329. Inappropriate reactions	
		43. Probing incorrect answers with reluctant learners How will I probe incorrect answers with reluctant learners?	330. Using an appropriate response process 331. Letting students off the hook temporarily 332. Using answer revision 333. Using think-pair-share	

Appendix B
List of Figures and Tables

Visit **go.SolutionTree.com/instruction** for free reproducible versions of figures and tables with an asterisk.

Figure I.1: The teaching and learning progression

Table I.1: Teacher Actions and Student Mental States and Processes

Table I.2: Design Questions

Table I.3: Elements Within the Ten Design Areas

Figure I.2: General format of the self-rating scale

Figure 1.1: Self-rating scale for element 1—Providing scales and rubrics

Table 1:1: The Three Dimensions of the Framework

Figure 1.2: Middle school life science performance expectation

Figure 1.3: Middle school science proficiency scale

Figure 1.4: Proficiency scale and success criteria for constructing scientific explanations

Figure 1.5: Self-rating scale for element 2—Tracking student progress

Figure 1.6: Student progress over time

Figure 1.7: Self-rating scale for element 3—Celebrating success

Figure 2.1: Self-rating scale for element 4—Using informal assessments of the whole class

Figure 2.2: Confidence levels progress chart*

Figure 2.3: Density triangle

Figure 2.4: Self-rating scale for element 5—Using formal assessments of individual students

Figure 2.5: Example proficiency scale

Figure 3.1: Self-rating scale for element 6—Chunking content

Figure 3.2: Accessing prior knowledge about water

Figure 3.3: Water cycle diagram

Figure 3.4: Biogeochemical cycle puzzle pieces

Figure 3.5: Biogeochemical cycle categorization grid

Figure 3.6: Self-rating scale for element 7—Processing content

Figure 3.7: Perspective analysis examples

Figure 3.8: Thinking hats example

Figure 3.9: Operons comparison chart example

Figure 3.10: Multiple-matching activity example

Figure 3.11: Self-rating scale for element 8—Recording and representing content

Figure 3.12: Teacher example of a tweet slide

Figure 3.13: Free-flowing web example of DNA replication

Figure 3.14: AP biology assignment example

Figure 4.1: Self-rating scale for element 9—Using structured practice sessions

Figure 4.2: Model for microscopy lesson

Figure 4.3: Self-rating scale for element 10—Examining similarities and differences

Figure 4.4: Self-rating scale for element 11—Examining errors in reasoning

Table 4.1: Examining Errors in Reasoning

Figure 5.1: Self-rating scale for element 12—Engaging students in cognitively complex tasks

Figure 5.2: Self-rating scale for element 13—Providing resources and guidance

Figure 5.3: Self-rating scale for element 14—Generating and defending claims

Figure 5.4: Before-and-after activity example

Figure 6.1: Self-rating scale for element 15—Previewing strategies

Figure 6.2: Self-rating scale for element 16—Highlighting critical information

Figure 6.3: Self-rating scale for element 17—Reviewing content

Figure 6.4: Performance expectations for a learning progression

Figure 6.5: Self-rating scale for element 18—Revising knowledge

Figure 6.6: Claim, evidence, and reasoning chart*

Figure 6.7: Self-rating scale for element 19—Reflecting on learning

Figure 6.8: Self-rating scale for element 20—Assigning purposeful homework

Figure 6.9: Self-rating scale for element 21—Elaborating on information

Figure 6.10: Self-rating scale for element 22—Organizing students to interact

Figure 6.11: Scaled target for constructing scientific explanations*

Figure 6.12: Making meaning and connections to reflect on learning

Figure 7.1: Self-rating scale for element 23—Noticing and reacting when students are not engaged

Figure 7.2: Self-rating scale for element 24—Increasing response rates

Figure 7.3: Self-rating scale for element 25—Using physical movement

Figure 7.4: Self-rating scale for element 26—Maintaining a lively pace

Figure 7.5: Self-rating scale for element 27—Demonstrating intensity and enthusiasm

Figure 7.6: Candle problem materials

Figure 7.7: Candle problem solution

Figure 7.8: Self-rating scale for element 28—Presenting unusual information

Figure 7.9: Teacher-created slide

Figure 7.10: History files activity example—Discovery of DNA

Figure 7.11: Student note-taking template

Figure 7.12: Self-rating scale for element 29—Using friendly controversy

Figure 7.13: Three ideas related to essential question about vegetarian diet

Figure 7.14: Venn diagrams to identify similarities and differences

Figure 7.15: Venn diagram comparing biological sex and gender

Figure 7.16: Self-rating scale for element 30—Using academic games

Figure 7.17: Which one doesn't belong? example

Figure 7.18: *Jeopardy!*-style gameboard

Figure 7.19: Self-rating scale for element 31—Providing opportunities for students to talk about themselves

Figure 7.20: All About Me activity example

Figure 7.21: Self-rating scale for element 32—Motivating and inspiring students

Figure 8.1: Self-rating scale for element 34—Organizing the physical layout of the classroom

Figure 8.2: Self-rating scale for element 39—Understanding students' backgrounds and interests

Figure 8.3: Student survey*

Figure 9.1: Developmental scale for elements

Figure 9.2: Self-audit for *The New Art and Science of Teaching* framework*

Figure 9.3: Teacher progress for the strategy of using physical movement—Body representation

Figure A.1: *The New Art and Science of Teaching* framework overview

References and Resources

Achieve Inc. (2013). *Next Generation Science Standards*. Washington, DC: Achieve Inc.

Achor, S. (2010). *The happiness advantage: The seven principles of positive psychology that fuel success and performance at work*. New York: Broadway Books.

Ainsworth, S. E., & Scheiter, K. (2021). Learning by drawing visual representations: Potential, purposes, and practical implications. *Current Directions in Psychological Science, 30*(1), 61–67.

Anderson, D. L., & Fisher, K. M. (n.d.). *Evolution concept cartoons with ?'s: Decide who is correct?* [PowerPoint slides]. SlidePlayer. Accessed at https://slideplayer.com/slide/15855678 on December 14, 2021.

Anderson, D. L., Fisher, K. M., & Norman, G. J. (2002). Development and evaluation of the conceptual inventory of natural selection. *Journal of Research in Science Teaching, 39*(10), 952–978.

Anderson, J. (2019). Cooperative learning: Principles and practice. *English Teaching Professional, 121,* 4–6.

Bamber, J., & Crowther, J. (2012). Speaking Habermas to Gramsci: Implications for the vocational preparation of community educators. *Studies in Philosophy and Education, 31*(2), 183–197.

Bamiro, A. O. (2015). Effects of guided discovery and think-pair-share strategies on secondary school students' achievement in chemistry. *SAGE Open, 5*(1), 1–7.

Bani-Salameh, H. N. (2016). How persistent are the misconceptions about force and motion held by college students? *Physics Education, 52*(1), 014003.

Blackburn, B. (2018). Productive struggle is a learner's sweet spot. *Productive Struggle for All, 14*(11). Accessed at www.ascd .org/ascd-express/vol14/num11/productive-struggle-is-a-learners-sweet-spot.aspx on February 3, 2021.

Bostan, C. M., Stanciu, T., & Andronic, R.-L. (2021). The moderation role of being valued by teachers over the association between school children motivation and need for competition. *SAGE Open, 11*(3), 1–9.

Brain Balance. (2022). *Normal attention span expectations by age*. Accessed at www.brainbalancecenters.com/blog/normal -attention-span-expectations-by-age on April 15, 2022.

BrainyQuote. (n.d.). *Robert Green Ingersoll quotes*. Accessed at https://www.brainyquote.com/quotes/robert_green _ingersoll_122475 on June 21, 2022.

Brown, P. C., Roediger, H. L., & McDaniel, M. A. (2014). *Make it stick: The science of successful learning*. Cambridge, MA: Belknap Press of Harvard University Press.

Chappuis, J. (2015). *Seven strategies of assessment for learning* (2nd ed.). Boston: Pearson Education.

Collaborative for Academic, Social, and Emotional Learning. (n.d.). *SEL: What are the core competence areas and where are they promoted?* Accessed at https://casel.org/fundamentals-of-sel/what-is-the-casel-framework on February 25, 2022.

Cooper, D. (2022). *Rebooting assessment: A practical guide for balancing conversations, performances, and products*. Bloomington, IN: Solution Tree Press.

de Bono, E. (1999). *Six thinking hats.* Boston: Back Bay Books.

Dean, C. B., Hubbell, E. R., Pitler, H., & Stone, B. J. (2012). *Classroom instruction that works: Research-based strategies for increasing student achievement* (2nd ed.). Alexandria, VA: ASCD.

Duncker, K. (1945). On problem-solving (L. S. Lees, Trans.). *Psychological Monographs, 58*(5), i–113.

Dunlosky, J., Rawson, K. A., Marsh, E. J., Nathan, M. J., & Willingham, D. T. (2013). Improving students' learning with effective learning techniques: Promising directions from cognitive and educational psychology. *Psychological Science in the Public Interest, 14*(1), 4–58.

Durlak, J. A., Weissberg, R. P., Dymnicki, A. B., Taylor, R. D., & Schellinger, K. B. (2011). The impact of enhancing students' social and emotional learning: A meta-analysis of school-based universal interventions. *Child Development, 82*(1), 405–432. Accessed at https://doi.org/10.1111/j.1467-8624.2010.01564.x on December 14, 2021.

Dweck, C. S. (2016). *Mindset: The new psychology of success* (Updated ed.). New York: Ballantine Books.

Ellis, S., Carette, B., Anseel, F., & Lievens, F. (2014). Systematic reflection: Implications for learning from failures and successes. *Current Directions in Psychological Science, 23*(1), 67–72.

Emmons, R. A., & McCullough, M. E. (2003). Counting blessings versus burdens: An experimental investigation of gratitude and subjective well-being in daily life. *Journal of Personality and Social Psychology, 84*(2), 377–389.

Ericsson, K. A., & Charness, N. (1994). Expert performance: Its structure and acquisition. *American Psychologist, 49*(8), 725–747.

Ericsson, K. A., Krampe, R. T., & Tesch-Römer, C. (1993). The role of deliberate practice in the acquisition of expert performance. *Psychological Review, 100*(3), 363–406.

Estrella, G., Au, J., Jaeggi, S. M., & Collins, P. (2018). Is inquiry science instruction effective for English language learners? A meta-analytic review. *AERA Open, 4*(2), 1–23.

Fry, R., Kennedy, B., & Funk, C. (2021, April). *STEM jobs see uneven progress in increasing gender, racial and ethnic diversity.* Accessed at www.pewresearch.org/science/2021/04/01/stem-jobs-see-uneven-progress-in-increasing-gender-racial-and-ethnic-diversity on April 13, 2022.

Gewertz, C. (2020, March 5). *The art of making science accessible and relevant to all students.* Accessed at www.edweek.org/teaching-learning/the-art-of-making-science-accessible-and-relevant-to-all-students/2020/03 on April 22, 2022.

Gijbels, D., Dochy, F., Van den Bossche, P., & Segers, M. (2005). Effects of problem-based learning: A meta-analysis from the angle of assessment. *Review of Educational Research, 75*(1), 27–61.

Glass, K. T., & Marzano, R. J. (2018). *The new art and science of teaching writing.* Bloomington, IN: Solution Tree Press.

Glenn Research Center. (n.d.). *Newton's laws of motion.* Accessed at www1.grc.nasa.gov/beginners-guide-to-aeronautics/newtons-laws-of-motion on April 20, 2021.

Harmon-Jones, E., Harmon-Jones, C., & Levy, N. (2015). An action-based model of cognitive-dissonance processes. *Current Directions in Psychological Science, 24*(3), 184–189.

Hattie, J. (2009). *Visible learning: A synthesis of over 800 meta-analyses relating to achievement.* New York: Routledge.

Hattie, J. (2017). *Hattie ranking: 252 influences and effect sizes related to student achievement.* Accessed at https://visible-learning.org/hattie-ranking-influences-effect-sizes-learning-achievement on March 6, 2022.

Hattie, J., & Timperley, H. (2007). The power of feedback. *Review of Educational Research, 77*(1), 81–112.

Haydon, T., Hunter, W., & Scott, T. M. (2019). Active supervision: Preventing behavioral problems before they occur. *Beyond Behavior, 28*(1), 29–35.

Heaslip, G., Donovan, P., & Cullen, J. G. (2013). Student response systems and learner engagement in large classes. *Active Learning in Higher Education, 15*(1), 11–24.

Heath, C., & Heath, D. (2017). *The power of moments: Why certain experiences have extraordinary impact.* New York: Simon & Schuster.

Howard, J. L., Bureau, J., Guay, F., Chong, J. X. Y., & Ryan, R. M. (2021). Student motivation and associated outcomes: A meta-analysis from self-determination theory. *Perspectives on Psychological Science, 16*(6), 1300–1323.

Kagan, S., & Kagan, M. (2015). *Kagan cooperative learning.* San Clemente, CA: Kagan.

Keeley, P. (n.d.). *Uncovering student ideas in science series.* Accessed at www.nsta.org/book-series/uncovering-student-ideas-science on December 28, 2021.

Keeley, P. (2011). *Uncovering student ideas in life science, Volume 1: 25 new formative assessment probes.* Arlington, VA: NSTA Press.

Keeley, P., & Cooper, S. (2019). *Uncovering student ideas in physical science, Volume 3:32 new matter and energy formative assessment probes.* Arlington, VA: NSTA Press.

Keeley, P., & Sneider, C. (2012). *Uncovering student ideas in astronomy: 45 formative assessment probes.* Arlington, VA: NSTA Press.

Kise, J. A. G. (2021). *Doable differentiation: Twelve strategies to meet the needs of all learners.* Bloomington, IN: Solution Tree Press.

Koch, I., Philipp, A. M., & Gade, M. (2006). Chunking in task sequences modulates task inhibition. *Psychological Science, 17*(4), 346–350.

Lang, J. M. (2021). *Small teaching: Everyday lessons from the science of learning* (2nd ed.). San Francisco: Jossey-Bass.

Lee, B. K., Patall, E. A., Cawthon, S. W., & Steingut, R. R. (2015). The effect of drama-based pedagogy on preK–16 outcomes: A meta-analysis of research from 1985 to 2012. *Review of Educational Research, 85*(1), 3–49.

Macnamara, B. N., Hambrick, D. Z., & Oswald, F. L. (2014). Deliberate practice and performance in music, games, sports, education, and professions: A meta-analysis. *Psychological Science, 25*(8), 1608–1618.

Maroszek, L. (n.d.). *Welcome to Mrs. Maroszek's website.* Accessed at http://maroszek.weebly.com/motion.html on December 14, 2021.

Martin, R., & Murtagh, E. M. (2017). Effect of active lessons on physical activity, academic, and health outcomes: A systematic review. *Research Quarterly for Exercise and Sport, 88*(2), 149–168.

Marzano, R. J. (2006). *Classroom assessment and grading that work.* Alexandria, VA: ASCD.

Marzano, R. J. (2007). *The art and science of teaching: A comprehensive framework for effective instruction.* Alexandria, VA: ASCD.

Marzano, R. J. (2010). Developing expert teachers. In R. J. Marzano (Ed.), *On excellence in teaching* (pp. 213–246). Bloomington, IN: Solution Tree Press.

Marzano, R. J. (2012). Art and science of teaching / teaching argument. *Educational Leadership, 70*(1), 80–81. Accessed at www1.ascd.org/publications/educational-leadership/sept12/vol70/num01/Teaching-Argument.aspx on December 23, 2021.

Marzano, R. J. (2016). *Direct instruction lessons: Processing content.* Bloomington, IN: Marzano Resources.

Marzano, R. J. (2017). *The new art and science of teaching.* Bloomington, IN: Solution Tree Press.

Marzano, R. J., & Eaker, R. (Eds.; 2020). *Professional Learning Communities at Work® and High-Reliability Schools™: Cultures of continuous learning.* Bloomington, IN: Solution Tree Press.

Marzano, R. J., Marzano, J. S., & Pickering, D. J. (2003). *Classroom management that works: Research-based strategies for every teacher.* Alexandria, VA: ASCD.

Marzano, R. J., Pickering, D. J., & Heflebower, T. (2011). *The highly engaged classroom.* Bloomington, IN: Marzano Resources.

Marzano, R. J., Yanoski, D. C., & Paynter, D. E. (2016). *Proficiency scales for the new science standards: A framework for science instruction and assessment.* Bloomington, IN: Marzano Resources.

McTighe, J., & Wiggins, G. (2013). *Essential questions: Opening doors to student understanding.* Alexandria, VA: ASCD.

Merriam-Webster. (n.d.). *Operon.* Accessed at www.merriam-webster.com/dictionary/operon on April 22, 2022.

Millar, R., Lubben, F., Got, R., & Duggan, S. (1994). Investigating in the school science laboratory: Conceptual and procedural knowledge and their influence on performance. *Research Papers in Education, 9*(2), 207–248.

Munroe, R. (2014). *What if? Serious scientific answers to absurd hypothetical questions.* Boston: Houghton Mifflin Harcourt.

Nagro, S. A., Hooks, S. D., Fraser, D. W., & Cornelius, K. E. (2016). Whole-group response strategies to promote student engagement in inclusive classrooms. *TEACHING Exceptional Children, 48*(5), 243–249.

National Research Council. (2012). *A framework for K–12 science education: Practices, crosscutting concepts, and core ideas.* Washington, DC: The National Academies Press.

Next Generation Science Standards. (n.d.a). *Evidence statements.* Accessed at www.nextgenscience.org/evidence-statements on December 14, 2021.

Next Generation Science Standards. (n.d.b). *MS-LS1-6 from molecules to organisms: Structures and processes.* Accessed at www.nextgenscience.org/pe/ms-ls1-6-molecules-organisms-structures-and-processes on May 29, 2019.

Next Generation Science Standards. (2013). *Appendix F: Science and engineering practices in the NGSS.* Accessed at www.nextgenscience.org/sites/default/files/Appendix%20F%20%20Science%20and%20Engineering%20Practices%20in%20the%20NGSS%20-%20FINAL%20060513.pdf on December 14, 2021.

NGSS Lead States. (2013). *Next Generation Science Standards: For states, by states.* Washington, DC: The National Academies Press.

Nye, B., Konstantopoulos, S., & Hedges, L. V. (2004). How large are teacher effects? *Educational Evaluation and Policy Analysis, 26*(3), 237–257.

Osborne, J. (2014). Teaching scientific practices: Meeting the challenge of change. *Journal of Science Teacher Education, 25*(2), 177–196.

Pink, D. H. (2009). *Drive: The surprising truth about what motivates us.* New York: Riverhead Books.

ProCon. (2021, February 25). *Should people become vegetarian?* Accessed at http://vegetarian.procon.org on December 14, 2021.

Rau, M. A. (2018). Making connections among multiple visual representations: How do sense-making skills and perceptual fluency relate to learning of chemistry knowledge? *Instructional Science, 46*(2), 209–243.

Reibel, A. R., & Twadell, E. (Eds.). (2019). *Proficiency-based grading in the content areas: Insights and key questions for secondary schools.* Bloomington, IN: Solution Tree Press.

Ritchart, R., Church, M., & Morrison, K. (2011). *Making thinking visible: How to promote engagement, understanding, and independence for all learners.* San Francisco: Jossey-Bass.

Savage, A., Hyneman, J., Wolkovitch, L., et al. (Executive Producers). (2003–2018). *MythBusters* [TV series]. Beyond Television Productions; Discovery Channel; Science Channel.

ScienceDirect. (n.d.). *Krebs cycle.* Accessed at www.sciencedirect.com/topics/engineering/krebs-cycle on April 13, 2022.

Silvia, P. J. (2008). Interest: The curious emotion. *Current Directions in Psychological Science, 17*(1), 57–60.

Smaglik, P. (2014, July). Media consulting: Entertaining science. *Nature, 511,* 113–115. Accessed at https://doi.org/10.1038/nj7507-113a on April 13, 2022.

ten Berge, T., & van Hezewijk, R. (1999). Procedural and declarative knowledge: An evolutionary perspective. *Theory & Psychology, 9*(5), 605–624.

Tomlinson, C. A., Brighton, C., Hertberg, H., Callahan, C. M., Moon, T. R., Brimijoin, K., et al. (2003). Differentiating instruction in response to student readiness, interest, and learning profile in academically diverse classrooms: A review of literature. *Journal for the Education of the Gifted, 27*(2–3), 119–145.

Walker, A. R., Collins, T. S., & Moody, A. K. (2014). Focus on family: Homework supports for children with learning disabilities. In S. Catapano (Ed.), *Childhood Education, 90*(4), 319–322.

Webb, N. M., & Mastergeorge, A. M. (2003). The development of students' helping behavior and learning in peer-directed small groups. *Cognition and Instruction, 21*(4), 361–428.

Wikimedia Commons. (2010, June 3). *Water cycle—blank.svg.* Accessed at https://commons.wikimedia.org/wiki/File:Water_Cycle_-_blank.svg on December 28, 2021.

Wiliam, D. (2018). *Embedded formative assessment* (2nd ed.). Bloomington, IN: Solution Tree Press.

Index

A

academic games, using, 121–124
Achor, S., 128
All About Me projects, 125, 126
assessments. *See also* preassessments
    about, 23
    designing assessments that generate formative scores, 17
    formal assessments of individual students, 27–30
    guiding questions for curriculum design for, 30
    informal assessments of the whole class, 23–27
    postassessments, 90, 106–107
    summary, 30
assignment revision, 86–87
attention, 103, 110–111

B

background knowledge, 35. *See also* prior knowledge
balance, use of, 56
before-and-after activity, 72, 73
biogeochemical cycle, examples for, 37
biology
    AP biology assignment example, 48–50
    dichotomous keys and, 60
    drama-related activities and, 109
    dramatic enactments for, 48
    mnemonic devices and, 51
    presenting unusual information and, 115, 116
    race and, 137
    response chaining and, 107
    teaching the water cycle, 35–36
    using friendly controversy and, 119–120
body representations, 108–109. *See also* physical
        movement, using
brain breaks, 105
building relationships. *See* implementing rules and
        procedures and building relationships

C

candle observations/problem, 111, 112–113
cartoon analysis, 42
catch and fix the error, 58
celebrating success, 19–21
Chappuis, J., 20
chemistry
    cumulative review and, 84
    hand signals and, 106
    Jeopardy!-style gameboard for, 123
    mnemonic devices and, 51
    personal stories and, 114
    preassessments and, 35
    summaries and, 45
    think-pair-share and, 95
    varied practice and, 58
choice, varied practice and, 57–58
chunking content, 34–38
circulating around the room, 70–71
claims, generating and defending, 72–74
classroom layout, organizing, 131–134
cognitive dissonance, creating, 71–72
cognitive processing styles, 127
cognitively complex tasks, 67–69
common assessments designed using proficiency scales, 28.
        See also assessments
comparisons
    comparison charts, 42
    constructed-response comparisons, 60
    diagrams comparing perspectives, 119–120
    knowledge comparisons, 90
con jobs, 111
concept attainment, 42–43
concept generalizations, 87. *See also* writing tools
concept webs, 46
conducting direct instruction lessons. *See* direct
        instruction lessons

conducting knowledge application lessons. *See* knowledge application lessons
conducting practicing and deepening lessons. *See* lessons, practicing and deepening
confidence rating techniques, 24–25
construct explanations, 59
constructed-response comparisons, 60
constructed-response items, 29
content. *See also* direct instruction lessons; knowledge application lessons; lessons, practicing and deepening; strategies that appear in all types of lessons
    about, 2
    design questions, 3–4
    elements within the design areas, 5
    reviewing content, 82–85
context, 2, 3, 4, 5, 103. *See also* engagement strategies; expertise, developing; implementing rules and procedures and building relationships
continuous improvement, 146
critical information, 80–82
crosscutting concepts (CCCs), 13, 98
cultural responsiveness, 19, 137
cumulative review, 83–84
curiosity creators, 111

D
declarative knowledge, 53, 59, 65
deliberate practice, engaging in, 144–145
demonstrations
    demonstrating intensity and enthusiasm, 112–114
    informational hooks and, 79
    reviewing content and, 84–85
density, 26, 83
design areas, 2–5. *See also specific design areas*
diagrams comparing perspectives, 119–120
dichotomous keys, 60–61
dimensions of learning framework, 12–13
direct instruction lessons
    about, 33
    chunking content, 34–38
    guiding questions for curriculum design for, 52
    processing content, 38–43
    recording and representing content, 44–51
    summary, 52
disciplinary core ideas (DCI), 13, 15, 67
diversity, 19, 137
DNA, examples with, 47, 51, 116
drama-related activities, 109
dramatic enactments, 48, 50
dramatic instruction, 82

E
egg-drop activity, 69
elaborating on information, 91–93
elements within design areas, list of 4, 5

elevator pitches, 45, 138
Embedded Formative Assessment (Wiliam), 17
energy level, 103. *See also* engagement strategies
engagement, use of term, 103
engagement strategies
    about, 103–104
    academic games, using, 121–124
    allowing students to talk about themselves, 124–127
    friendly controversy, using, 118–120
    guiding questions for curriculum design for, 128–130
    information, presenting unusual, 114–118
    intensity and enthusiasm, demonstrating, 112–114
    motivating and inspiring students, 127–128
    noticing and reacting when students are not engaged, 104–105
    pacing, maintaining lively, 109–112
    physical movement, using, 108–109
    response rates, increasing, 105–107
    summary, 130
errors, catch and fix, 58
errors in reasoning, examining, 62–64
essential questions, 38
evidence, 72–74, 88
exit slips, 89
experimental designs, 82
experimental-inquiry tasks, 68–69
expertise, developing, 141–146

F
fast facts, 114–115
feedback
    about, 2
    assignment revision and, 87
    characteristics of/offering feedback, 71
    design questions, 3
    elements within the design areas, 5
    observations of students and, 30
    peer feedback, 86
    providing and communicating clear learning goals. *See* learning goals
    using assessments. *See* assessments
    verbal feedback, 20–21
fist to five, 24
fixed mindset, 20. *See also* growth mindset
flubber activities, 71
force concept inventory (FCI), 90
formal assessments of individual students, 27–30. *See also* assessments
formative assessments, 17. *See also* assessments
free-flowing webs, 46–47
friendly controversies, using, 118–120
future growth, planning for, 146

G
gas laws, 79, 106
give one, get one, 85

goals, selecting goal elements and specific strategies, 144. *See also* learning goals
graphical modes, 46
gratitude journals, 128
grounds, providing, 72–74
group work, classroom organization for, 134
grouping students, 93–98
growth mindset, 20, 87. *See also* fixed mindset
guest speakers, 116, 118
guided practice, 56–57

H
hand signals, 105–107
Happiness Advantage, The (Achor), 128
hashtags, 45
Hedges, L., 141
help-giving/help-receiving behavior, 85
heterogeneous grouping, 93, 94. *See also* grouping students
highlighting critical information, 80–82
history files, 116
homework, 90–91
homogenous grouping, 93, 94. *See also* grouping students
hooks, 78–79, 111–112
huddleboards. *See* whiteboards, use of

I
implementing rules and procedures and building relationships
    about, 131
    guiding questions for curriculum design for, 138–139
    organizing the physical layout of the classroom, 131–134
    summary, 139
    understanding students' backgrounds and interests, 134–138
inconsequential competitions, 122–123
increasing response rates, 105–107
informal assessments of the whole class, 23–27. *See also* assessments
information, presenting unusual, 114–118
informational hooks, 78–79
inquiry-based activity, 70–71
instructional segments, 110–111
intensity and enthusiasm, demonstrating, 112–114
interest surveys, 125
interrogations, elaborative, 92–93
intrigue, 103. *See also* engagement strategies
introduction
    about series, 1–2
    about this book, 7–8
    need for subject-specific models, 6–7
    overall model, 2–6
    invention tasks, 69

J
Jeopardy!-style games, 122–123

K
Kahoot! 122, 123
kinematics, 84, 85
knowledge application lessons
    about, 67
    claims, generating and defending, 72–74
    cognitively complex tasks, engaging students in, 67–69
    guiding questions for curriculum design for, 74–75
    resources and guidance, providing, 69–72
    summary, 75
knowledge comparisons, 90
knowledge gain celebrations, 19
Konstantopoulos, S., 141
Krebs cycle, 35

L
lab activities/skills
    and classroom materials, computers, and technology equipment, 133–134
    and elaborative interrogations, 92
    and grouping, 94
    and modeling, 54
    and observations of students, 29–30
    and think-pair-share and think-pair-square, 95
    and visual activities, 82
Lang, J., 80
learning goals
    about, 11
    celebrating success, 19–21
    guiding questions for curriculum design for, 21–22
    providing scales and rubrics, 11–16
    summary, 22
    tracking student progress, 16–18
learning progressions, 1, 16, 84
learning targets, 11, 16, 96–97
lessons, direct instruction. *See* direct instruction lessons
lessons, knowledge and application. *See* knowledge application lessons
lessons, practicing and deepening
    about, 53
    examining errors in reasoning, 62–64
    examining similarities and differences, 59–62
    guiding questions for curriculum design for, 64–65
    summary, 65
    using structured practice sessions, 54–59

M
measurements, 84, 91
media, finding errors in, 64
minute thesis, 97–98
misinformation, identifying errors of, 62–63
mnemonic devices, 51
model of *The New Art and Science of Teaching*
    about, 2–6, 147
    need for subject-specific models, 6–7

The New Art and Science of Teaching framework overview, 150–161
modeling, 54–56
mood meters, 104–105
motivating and inspiring students, 103, 127–128. *See also* engagement strategies
motivational hooks, 111–112
multiple-choice questions, 29, 106, 123
multiple-matching activities, 42–43
mystery draw, 111–112
myth busting, 79

N
nonlinguistic representations, 112–113
norms, creating group norms, 94
note-taking, template for, 117
noticing and reacting when students are not engaged, 104–105
Nye, B., 141

O
observations of students, 29–30
offering feedback, 71
operon, 42, 43
opinion questionnaires, 136–137
organelle speed dating, 115
organizing students to interact, 93–98

P
pacing, maintaining lively, 109–112
peer feedback, 20, 86
peer response groups, 95
performance expectations, 13, 14, 17, 18, 84
personal stories, 113–114
perspective analysis, 39–40
physical movement, using, 108–109
physics
    body representations and, 109
    cumulative review and, 84
    knowledge comparisons and, 90
    media, finding errors in, 64
    summaries and, 45
pinhole camera, 78
planning for future growth, 146
polling, 27
possible selves activities, 128
postassessments, 90, 106–107. *See also* assessments
practicing and deepening lessons. *See* lessons, practicing and deepening
preassessments. *See also* assessments
    chunking content and, 34–35
    grouping using, 94
    hand signals and, 106–107
    preview questions and, 80
predictions, 80
preview questions, 80

previewing strategies, 78–80
prior knowledge
    chunking content and, 35, 36
    highlighting critical information and, 81
    mnemonic devices and, 51
    previewing strategies and, 78, 79–80
    reviewing content and, 83
problem-based learning, 84
procedural knowledge versus declarative knowledge, 53
processing content, 38–43
processing time, 38
proficiency scales
    common assessments designed using, 28
    confidence rating techniques and, 24
    example proficiency scale, 15, 16, 28
    formative assessments and, 17
    knowledge gain celebrations, 19
    learning goals and, 11
    observations of students and, 30
    providing scales and rubrics, 11–16
    tracking student progress and, 16
progress monitoring
    confidence levels progress chart, 25
    deliberate practice and, 144–145
    tracking student progress, 16–18
    verbal feedback and, 20
providing and communicating clear learning goals. *See* learning goals
providing opportunities for students to talk about themselves, 124–127

Q
questions/questionnaires
    elaborative interrogations, 92–93
    experimental-inquiry tasks, 68
    multiple-choice questions, 29, 106, 123
    opinion questionnaires, 136–137
    preview questions, 80
    question prompts, 57
Quizizz, 27, 122, 123

R
Rally Coach strategy, 56
recording and representing content
    about, 44
    dramatic enactments, 48, 50
    free-flowing webs, 46–47
    mnemonic devices, 51
    summaries, 45–46
re-engaging individual students, 104–105
reflecting on learning (groups), 97–98
reflecting on learning (individual), 20, 88–90
relationships, 2, 21. *See also* implementing rules and procedures and building relationships
repetition, 81
resources and guidance, providing, 69–72

response boards, 26–27
response cards, 107
response chaining, 107
response rates, increasing, 105–107
reviewing content, 82–85
revising knowledge, 85–88
rules and procedures. *See* implementing rules and
    procedures and building relationships

S
scales and rubrics, providing, 11–16
science and engineering practices (SEPs), 98
SEL (social-emotional learning), 20–21
selected response, 29
self-audit, conducting, 142, 143
self-rating scales, general format of, 7. *See also
    specific elements*
seminars, 119
sentence starters, 57
sets, presenting content in, 35–38
similarities and differences, examining, 59–62
single-word summaries, 46
six-word autobiographies, 138
Small Teaching (Lang), 80
social-emotional learning (SEL), 20–21
sorting, matching, and categorizing, 61–62
Stand Up, Hand Up, Pair Up, 85
stems, 87–88. *See also* writing tools
strategies that appear in all types of lessons
    about, 77
    assigning purposeful homework, 90–91
    elaborating on information, 91–93
    guiding questions for curriculum design for, 99–100
    highlighting critical information, 80–82
    organizing students to interact, 93–98
    previewing strategies, 78–80
    reflecting on learning, 88–90
    reviewing content, 82–85
    revising knowledge, 85–88
    summary, 100
structured practice sessions, 54–59
student background surveys, 135–136
student learning profiles, 125–127
student progress, charting, 18
student work, displaying, 132–133
students, allowing to talk about themselves, 124–127
students, motivating and inspiring, 127–128
students, re-engaging individual, 104–105
students' backgrounds and interests, 134–138
success criteria
    exit slips and, 89
    knowledge gain celebrations and, 19
    observations of students and, 29–30
    peer feedback and, 86
    peer response groups and, 95

proficiency scale and success criteria for constructing
    scientific explanations, 16
    verbal feedback and, 20
summaries, 45–46
surveys, interest, 125
surveys, student background, 135–136

T
teacher efficacy, impact of, 141
teacher impact on student learning, 141
teaching and learning progression, 1
technology, 133–134
thinking hats, 40–42
think-pair-share and think-pair-square, 95
three truths and a lie, 121
top-ten lists, 46
tracking student progress, 16–18, 144–145
turn taking, 57

U
unusual information. *See* information, presenting unusual
using assessments. *See* assessments
using strategies that appear in all types of lessons.
        See strategies that appear in all types of lessons

V
varied practice, 57–59
vegetarianism, 95, 119
Venn diagrams, 60, 119–120
verbal feedback, 20–21. *See also* feedback
visual activities, 82
vocabulary
    academic games and, 121
    displaying student work and, 133
    prior knowledge and, 36
    vocabulary review games, 124

W
water cycle, 35–36
what do you think you know? 79–80
which one doesn't belong? 121–122
whiteboards, use of, 26, 35, 38, 79, 98, 107, 111, 122, 133
whole class, informal assessments of, 23–27.
        See also assessments
Wiliam, D., 17
word webs, free-flowing webs, 46
writing tools, 87–88

Solution Tree

Solution Tree's mission is to advance the work of our authors. By working with the best researchers and educators worldwide, we strive to be the premier provider of innovative publishing, in-demand events, and inspired professional development designed to transform education to ensure that all students learn.

ASCD is a global nonprofit association dedicated to the whole child approach that supports educators, families, community members, and policy makers. We provide expert and innovative solutions to facilitate professional development through print and digital publishing, on-site learning services, and conferences and events that empower educators to support the success of each child.